INTERNATIONAL ENERGY AGENCY

SAVING OIL
IN
A HURRY

96811110

INTERNATIONAL ENERGY AGENCY

The International Energy Agency (IEA) is an autonomous body which was established in November 1974 within the framework of the Organisation for Economic Co-operation and Development (OECD) to implement an international energy programme.

It carries out a comprehensive programme of energy co-operation among twenty-six of the OECD's thirty member countries. The basic aims of the IEA are:

- to maintain and improve systems for coping with oil supply disruptions;
- to promote rational energy policies in a global context through co-operative relations with non-member countries, industry and international organisations;
- to operate a permanent information system on the international oil market;
- to improve the world's energy supply and demand structure by developing alternative energy sources and increasing the efficiency of energy use;
- to assist in the integration of environmental and energy policies.

The IEA member countries are: Australia, Austria, Belgium, Canada, the Czech Republic, Denmark, Finland, France, Germany, Greece, Hungary, Ireland, Italy, Japan, the Republic of Korea, Luxembourg, the Netherlands, New Zealand, Norway, Portugal, Spain, Sweden, Switzerland, Turkey, the United Kingdom, the United States. The European Commission takes part in the work of the IEA.

ORGANISATION FOR ECONOMIC CO-OPERATION AND DEVELOPMENT

The OECD is a unique forum where the governments of thirty democracies work together to address the economic, social and environmental challenges of globalisation. The OECD is also at the forefront of efforts to understand and to help governments respond to new developments and concerns, such as corporate governance, the information economy and the challenges of an ageing population. The Organisation provides a setting where governments can compare policy experiences, seek answers to common problems, identify good practice and work to co-ordinate domestic and international policies.

The OECD member countries are: Australia, Austria, Belgium, Canada, the Czech Republic, Denmark, Finland, France, Germany, Greece, Hungary, Iceland, Ireland, Italy, Japan, Korea, Luxembourg, Mexico, the Netherlands, New Zealand, Norway, Poland, Portugal, the Slovak Republic, Spain, Sweden, Switzerland, Turkey, the United Kingdom and the United States. The European Commission takes part in the work of the OECD.

FOREWORD

During 2004, oil prices reached levels unprecedented in recent years. Though world oil markets remain adequately supplied, high oil prices do reflect increasingly uncertain conditions. Many IEA member countries and non-member countries alike are concerned about oil costs and oil security and are looking for ways to improve their capability to handle market volatility and possible supply disruptions in the future. This book aims to provide assistance.

A core mission of the International Energy Agency (IEA) is energy supply security. The Agreement on an International Energy Program (IEP), the treaty signed by all IEA member countries, obliges them to maintain emergency oil reserves and be prepared to draw on these under certain circumstances, such as a sudden, substantial reduction in oil supplies to world markets. This agreement also requires development of voluntary and mandatory measures for rapidly reducing oil consumption ("demand restraint") under such circumstances. As the transport sector in most OECD countries is the prime consumer of oil, this sector should be a central focus of IEA member countries' emergency oil demand restraint programmes.

This book provides a new, quantitative assessment of the potential oil savings and costs of rapid oil demand restraint measures for transport. These measures may be useful both for large-scale supply disruptions – such as could lead to collective actions by IEA members – and for smaller or more localised supply disruptions in individual countries. Some measures make sense under any circumstances; others are primarily useful in emergency situations. A principal goal is to provide policy alternatives to measures like fuel rationing – never a good idea if it can be avoided.

Our estimates for each IEA region and for the IEA in total show that several types of demand restraint measures could provide large reductions in oil use quickly and cheaply, while helping preserve mobility options. But for these measures to succeed, countries must be well prepared and act aggressively during an emergency. The book also provides clear methodologies that individual countries can use to make their own estimates. IEA and non-IEA countries alike are encouraged to engage in such analysis and consider which policies would be best adapted to their national circumstances.

Perhaps most importantly, this book is intended to raise awareness that demand response is an important aspect in dealing with supply disruptions. Oil demand in transport is indeed very "inelastic" in the short run, but the set of measures outlined here gives countries a tool box they can draw on to help lower the duration and costs of petroleum supply disruptions.

Claude Mandil
Executive Director

ACKNOWLEDGEMENTS

This publication is the product of an IEA study jointly undertaken by the Divisions of Emergency Planning and Preparations (directed by Klaus Jacoby) and Energy Technology Policy (directed by Fridtjof Unander). Important direction was also provided by Carmen Difiglio, Marianne Haug and Neil Hirst. The study was co-ordinated by Lew Fulton, Enno Harks and Tom Howes.

This publication is based on a report prepared for the IEA by Robert Noland (Imperial College London) and William Cowart (ICF Consulting). The final manuscript was prepared by Lew Fulton with contributions from Noé van Hulst, Kenji Kobayashi, Pierre Lefèvre, Fatih Birol, Klaus Jacoby, Jason Elliott, Rebecca Gaghen, Thomas Guéret, James Haywood, Kristine Kuolt, Alan Meier and James Ryder.

The IEA would like to express its appreciation to all those who attended the workshop "Managing Oil Demand in Transport" (7-8 March 2005) and contributed ideas and suggestions that are reflected in this publication. Special thanks to Heather Allen (International Union of Public Transport, Brussels); David L. Greene (Oak Ridge National Laboratory, US); Martin Kroon (Netherlands Ministry of Housing, Physical Planning and Environment); John Mumford (BP Oil UK Ltd); Heather Staley (Energy Efficiency and Conservation Authority, New Zealand) and Jack Short and Philip Watson (both of OECD/ECMT, Paris).

Assistance with editing and preparation of the manuscript was provided by Viviane Consoli, Muriel Custodio, Corinne Hayworth, Janet Pape and Bertrand Sadin.

TABLE OF CONTENTS

LIST OF FIGURES

LIST OF BOXES

LIST OF TABLES

EXECUTIVE SUMMARY

In October 1973, an embargo by several Middle Eastern countries caused oil supply shortages for several months in most IEA countries and many other countries around the world. Since then, disruptions affecting world oil supply and prices have occurred fairly regularly, averaging two to three significant episodes per decade. In each instance, supplies of retail fuel have gone into shortage in one or more countries and oil prices have risen rapidly and substantially.

To offset a large oil supply disruption, the IEA and its member countries have an important tool: emergency stocks and associated stock draw. In addition to this powerful measure, oil demand restraint measures in transport can be useful. The transport sector accounts for over half of oil use in IEA countries and is expected to account for nearly all future growth in oil use. This book explores measures to rapidly reduce oil demand in the passenger transport sector, over short periods of time.

Application of transportation "demand restraint" policies have increasingly been used by cities around the world to quickly reduce air pollution levels during periods of unacceptably bad air quality; a similar approach may also be useful in the event of oil supply disruptions or during periods of high oil prices. Some measures may also be attractive to cut oil demand over longer periods of time – for example during extended periods of high oil prices, to relieve demand pressure on the market or to rapidly cut the use of an expensive fuel. This is particularly true for measures that can reduce fuel use at low cost, while preserving mobility options. This book emphasises that there are important differences between measures that simply restrict travel, such as driving bans, and those that assist motorists to rapidly cut fuel use, such as promoting "ecodriving" and facilitating car-pooling.

Background and approach

There have been many previous studies of options to reduce oil use in transport. Usually such studies evaluate a range of policy options used under normal circumstances to manage transport fuel demand (or demand for transport itself) in order to reduce long-term growth and/or environmental impacts. While such analysis is very important, and the IEA urges all countries to pursue strategies for reducing oil use over the medium and long term, the analysis presented here differs in an important respect. It focuses on a much shorter time frame: the circumstances of a temporary oil supply disruption that may result in physical shortages of oil or a sudden severe price spike. As we show, this difference in time frame and circumstance can result in a quite different type of analysis, with different results, than in many previous longer-term studies.

Why should governments intervene?

Why should governments intervene to cut oil demand during a supply disruption? One obvious reason is to conserve fuel that might be in short supply. A rapid demand response can also send a strong market signal. In the case of a moderate reduction in oil supplies, a reduction in IEA transport fuel demand of even a few percent could have a substantial dampening effect on surging world oil prices. Achieving even this much reduction in transport energy use would be challenging but, if successful, the value to IEA and other oil-importing countries in terms of maintaining adequate supplies and moderating oil price spikes could be far greater than the costs associated with the measures themselves.

Of course, a supply disruption that induces a rise in oil prices will generate its own response from drivers and other travellers. However, short-term transport demand response to changes in fuel price is notoriously slow and small (*i.e.* there is a low "price elasticity" of demand). If governments can provide better travel alternatives and other incentives to rapidly cut the most energy-intensive types of travel (such as driving alone) during supply disruptions, the response rate might be much higher and the disruption-related costs to society much lower.

Some measures that may not be attractive as general transport demand policies may be more effective in the context of an oil supply disruption. A number of new measures emerge that have not previously received much attention. Some otherwise costly measures appear to become less expensive if implemented over a short period of time, provided governments have taken the necessary preparatory steps to be ready to act on short notice. Several measures are likely to be more socially and politically acceptable, and therefore easier to implement, during an emergency than under normal circumstances. On the other hand, some of the measures discussed in this volume would make perfect sense as part of a more general transport oil demand management strategy. There is certainly some overlap between the two contexts.

Assessing impacts of measures under emergency conditions

There are several ways in which behaviour under conditions of oil supply shortage may differ from behaviour under normal circumstances. For example, the immediate public reaction to a drop in fuel supplies may be panic and hoarding behaviour. In such cases, demand may increase even with a sharp increase in prices. During such a situation, it may be important to manage oil supplies very tightly – for example through an oil allocation scheme. However, as rationing is rarely an economically efficient solution, once the situation is under control, governments should generally try to move quickly to approaches that are likely to be less costly to society. A major cost associated with a fuel shortage that is not solved by rationing is lost mobility – and the lost economic activity that results. If societal mobility can be maintained (*e.g.* through increased car-pooling), and/or if good alternatives to mobility can be provided (such as telecommuting) so that economic activities can continue, this will yield a much better societal outcome than through rationing-oriented solutions.

An important finding of this book is that pre-planning is essential in order for transport demand restraint measures to succeed during an emergency. It is not enough for countries to have a list of measures to use; they must be ready to implement those measures on very short notice. To do this, they generally must develop detailed plans and make certain investments ahead of time. Communicating this plan to the public also appears very important; if the public is not well informed of plans ahead of time, and supportive of them, they may be less likely to co-operate and do their part to help the plans succeed during an emergency. Strong support and co-operation from the business community is also essential. In general, providing clear information to the public – that the public can trust – seems to be an important element of any plan. The role of information is stressed throughout the analysis of measures in this book.

Once measures are put in place that provide fuel-efficient mobility options or alternatives to travel, the public responsiveness to these measures may actually be better during an emergency than under normal circumstances, since there will likely be a strong interest in such alternatives. There may also be an altruistic attitude amongst people to "do their part" during the emergency. If this occurs, then estimates of policy response and impacts based on behaviour during normal circumstances, as made throughout this analysis, may underestimate the impacts of measures during emergencies. But the relative impacts and costs of different measures should at least be similar.

Scope and approach of the study

The analysis presented in this report covers measures affecting road transport (including public transit). It focuses primarily on urban passenger travel, though in a couple of the assessed measures, such as speed limit reductions on motorways, trucks are also affected. There may also be important opportunities to save oil quickly in other transport sectors and modes, such as air travel, shipping, etc., and these should be investigated as well. But this study focuses primarily on road passenger transport because it appears to have some particularly promising opportunities to save oil quickly, and because relatively good information is available upon which to build an analysis.

Most measures considered here are focused on urban or metropolitan areas and therefore not typically applied at a national level (such as increasing public transit service). However, national governments are best positioned to launch a comprehensive programme for dealing with emergency situations, which could include creating incentives and working with cities and regional governments to establish similar programmes around the country.

The basic approach has been to evaluate the impact of a variety of measures if applied individually during an emergency, given the necessary planning and preparation before an emergency occurs. In most cases the measures have the effect of reducing private vehicle travel, either by reducing travel demand or encouraging shifts to public transit or other modes. The following general approaches were evaluated:

- Increases in public transit usage
- Increases in car-pooling
- Telecommuting (working at home)

- Changes in work schedules

- Driving bans and restrictions

- Speed limit reductions

- Information on "ecodriving" (*e.g.* driving style, tyre inflation)

Although most of these measures can also be used for more general transportation demand management, here they are assessed only in the context of temporary use during an emergency. Within each of these approaches several more specific measures were identified and evaluated. A representative measure was then selected with a "consensus" estimate of the likely effect. For example, for car-pooling, measures are assessed ranging from a simple policy of a public campaign calling on people to car-pool more, to actual improvements in car-pooling infrastructure (before an emergency occurs) and requirements that during the emergency cars carry more than one person on certain roads or for certain types of trips. Clearly, such a range of policy approaches can lead to a wide range of possible outcomes. We have provided estimates for many of these. In addition, for each policy type we have provided a consensus estimate based upon our judgment of most likely impacts.

Though driving bans are covered here, there are other types of rationing schemes that this analysis does not address, such as fuel allocation coupon systems. These types of measures may be needed, but should be seen as something of a last resort. Measures to reduce oil demand voluntarily appear likely to incur lower costs on society than simply restricting the supply of motor fuel. However, measures to reduce fuel "hoarding" and similar behaviours may provide an important complement to measures described here.

Policies aimed at changing the price of road transport, either through increased fuel taxes or road charging (toll fees), are discussed but not explicitly scored in terms of impacts. These types of policies, while capable of yielding reductions in fuel consumption, could be difficult to implement during a short-term emergency when fuel costs already may be rising rapidly. Oil price increases will likely have some (though perhaps small) dampening effect on transportation fuel demand, and help to spur the types of travel changes that this report focuses on, such as increased use of public transit, car-pooling and telecommuting. Therefore, perhaps the main price-related issue for policy-makers during a fuel supply disruption is to avoid bowing to pressure to lower existing fuel tax or road charging regimes, so that pricing signals are not distorted. In any case, the measures we focus on involve providing travellers with better information and alternatives to driving (especially to driving alone), so that their ability to respond to and cope with an oil emergency improves. Increased demand responsiveness reduces the negative economic impacts of a supply disruption.

In Chapter 2, estimates of the effects of different measures on oil demand are made for four IEA regions (Japan/Republic of Korea, IEA Europe, USA/Canada and Australia/New Zealand) and then summed over the whole IEA. Wherever possible, sources and data are used for specific countries within each region and aggregated to regional totals, with specific assumptions outlined for each measure. However, though this analysis is based upon existing estimates within the literature, there is a severe shortage of data covering the application of measures during emergency situations. Nor is there much quantitative evidence of how behaviour (such as responsiveness to policies) changes during an emergency. The transport literature generally analyses the longer-term effects associated with various policies under normal fuel supply conditions. Therefore, judgment has been used to estimate behaviour and responses to policies in such situations.

In some cases where data were not available, estimates from similar countries or regions have been used. The year 2001 was used as a "base year" for most calculations, since this was the most recent year for which enough data could be obtained to carry out detailed calculations. Though the amounts of driving and fuel consumption have changed since then, the relative impacts of different measures and the estimated percentage reductions should remain similar, for many years, to the results shown here. Much of the data used in the analysis is provided in tables throughout the report and in the appendix, in an effort to provide countries with much of the data they will need to conduct their own analyses.

Summary of results

A summary of our results, summed and averaged across all IEA countries, is shown in Table E-1. This table provides a brief overview of the types of strategies and the policy context needed to achieve these reductions. These estimates carry

Table E-1

Summary of oil-saving effects of measures summed across all IEA countries

Potential oil savings by category if implemented in all IEA countries	Measure
VERY LARGE More than one million barrels per day	**Car-pooling:** large programme to designate emergency car-pool lanes along all motorways, designate park-and-ride lots, inform public and match riders
	Driving ban: odd/even licence plate scheme. Provide police enforcement, appropriate information and signage
LARGE More than 500 thousand barrels per day	**Speed limits:** reduce highway speed limits to 90 kph. Provide police enforcement or speed cameras, appropriate information and signage
	Transit: free public transit (set fares to zero)
	Telecommuting: large programme, including active participation of businesses, public information on benefits of telecommuting, minor investments in needed infrastructure to facilitate
	Compressed work week (fewer but longer workdays): programme with employer participation and public information campaign
	Driving ban: 1 in 10 days based on licence plate, with police enforcement and signage
	"Ecodriving" (efficient driving styles and vehicle maintenance steps): intensive public information programme
MODERATE More than 100 thousand barrels per day	**Transit fare reduction:** 50% reduction in current public transit fares **Transit service increase:** increase weekend and off-peak transit service and increase peak service frequency by 10% **Car-pooling:** small programme to inform public, match riders
SMALL Less than 100 thousand barrels per day	**Bus priority:** convert all existing car-pool and bus lanes to 24-hour bus priority usage and convert some other lanes to bus-only lanes

a range of uncertainty in terms of the absolute value of the reductions which may be achieved. However, the orders of magnitude and relative effects between policies appear reasonable. The policy strategies shown are to a large degree mutually exclusive. Potential combinations of these measures have not been assessed. Clearly, a combined package of policies could increase the impacts compared to just one, but probably would not have an effect equal to the sum of these policies – since, for example, one person cannot both car-pool and telecommute on the same day. A proper analysis of mutual exclusivities and synergistic effects would require developing a detailed travel demand model and is beyond the scope of the methods used here to estimate these savings. However, more detailed approaches might be appropriate for individual countries – and are commonly available for large cities.

As shown in Table E-1, there is a large range of estimated effectiveness based upon both the specific strategy selected and the policy context in which it is pursued. In general, there are two types of policy approaches. One is focused on providing people with better (and less energy-intensive) travel options to allow them to save fuel, as well as allowing them to avoid the consequences of not being able to purchase fuel. These options tend to focus on providing people with more choices, such as better or cheaper public transit, car-pooling options, telecommuting, flexible work schedules, or promotion of "ecodriving" (efficient driving styles and vehicle maintenance steps). The other policy approach is more prohibitive in nature, essentially restricting travel options or requiring shifts in behaviour. These include driving bans, mandatory car-pooling, speed limit reductions or changes in work schedules. Not surprisingly, the more restrictive options tend to result in greater estimated reductions in fuel consumption, but may also be more expensive to society and unpopular and therefore less politically feasible.

Our main conclusions on those policies which can be most effective are as follows:

- **Restrictions on driving**, such as odd/even-day driving bans, can potentially provide very large savings. However, they may be unpopular and restrict mobility much more than some other measures. Multiple-vehicle households tend to be less affected by this type of policy and therefore this option may be seen as less equitable than some others. If conducted over longer periods, the effectiveness of such policies may decline as travellers figure out ways around the regulations.

- **Measures to increase car-pooling**, if successful, can provide rapid, large reductions in oil demand. But success may be very dependent on the level of incentives given to drivers, which could make this option quite costly. Restrictive options that require car-pooling (such as restricting certain traffic lanes to car-pools) are likely to be most effective but may be seen as inequitable, unless fairly limited in application. Programmes focused only on provision of information (such as setting up a web site to help potential car-poolers find other car-poolers) will likely be more popular, if less effective.

- **Reducing speed limits** on motorways can be very effective for saving fuel, since cars and trucks use much more fuel per kilometre as speeds increase above 90 kilometres per hour (about 55 mph). However, success depends on an adequate enforcement regime. In some cases better enforcement of existing speed limits may be sufficient to lower average speeds significantly. Clear information to the public regarding the strong links between lower speeds and fuel savings may help increase compliance during an emergency. An infrastructure allowing a rapid change to posted speed limits (such as variable speed limit signs) must be put in place ahead of time.

These types of policies, requiring some coercion or restriction on behaviour, may be more acceptable to the public during emergency situations than otherwise, if a sense of the need for common sacrifice is prevalent. In any case, popularity is likely to be fairly low.

In contrast, policies that provide mobility options, such as making it easier for people to use "alternative" modes (*i.e.* alternative to single-occupant vehicles), are likely to be popular, but have a range of effectiveness depending upon the measure and level of investment made. Some require significant investments in order to be prepared before an emergency occurs, so that implementation during an emergency can be achieved on a very short time scale.

- **Temporarily eliminating public transit fares** (*e.g.* if the lost transit revenues are covered out of general tax revenues) appears moderately effective, but would likely be relatively costly per barrel of oil saved. There would also be a large (and inefficient) windfall to existing riders. However, it may help increase the effectiveness and acceptance of other options such as driving bans.

- **Increasing transit service** can provide significant fuel savings (since this can cut car usage as travellers switch modes). However, for short-term increases using existing equipment and personnel, only a small expansion in services appears possible (*e.g.* peak hour services can be increased by perhaps 10% for most systems and peak services can be extended for longer time periods). Temporarily creating new bus-only lanes (or bus and car-pool-only lanes) by converting regular lanes can help, as can extending the operating hours of existing bus-only lanes.

A third set of policies can best be considered "no regret" policies. That is, they are likely to be relatively cheap to implement, mainly requiring a good public information campaign with some related support such as development of websites or other outreach programmes. While in some cases these will provide only modest oil savings, for an aggressive (and successful) programme the fuel savings could be quite large – up to one million barrels a day across all IEA countries. Public support for these measures is likely to be fairly good. Thus, these might be good measures to implement any time, on a permanent basis, though their impacts may be highest in an emergency situation, when the public is most likely to be responsive.

- **Telecommuting and flexible work schedules** can save substantial fuel and potentially be implemented very quickly. A well organised "emergency telecommuting" programme, particularly one where employers agree in advance to participate and designate certain employees to telecommute during designated situations, could yield large reductions in fuel use on such days. This type of plan could extend to other transportation-related emergencies, such as air quality "code red" days, transit strikes, etc.

- **Ecodriving** includes a wide array of behavioural changes, such as more efficient driving styles (*e.g.* changes in acceleration/deceleration and gear shifting patterns), optimal tyre inflation, reducing vehicle weight and other steps. An aggressive and comprehensive public information campaign on the benefits of "ecodriving" could yield substantial fuel savings. While some countries already run information campaigns of this type, at least occasionally, much stronger efforts could generate much better compliance, especially during emergencies.

Some other measures, such as switching to alternative fuels and improving new car fuel economy, were judged unlikely to have much impact in an emergency situation, when only very rapid reductions are useful. Measures in these areas may be very important over the medium and longer term, however, in order to lower the trend in transport oil use.

Regional differences

The estimated effectiveness of the different measures varies significantly between IEA regions. This is mainly due to variations in the transport sector in terms of mode shares and the resulting flexibility of travellers to change modes in each region. Figure ES-1 shows results for each region, for selected measures, as a percentage reduction in total petroleum fuel use for that region.

Figure E-1

Percentage reduction in total petroleum fuel use by IEA region, for selected measures

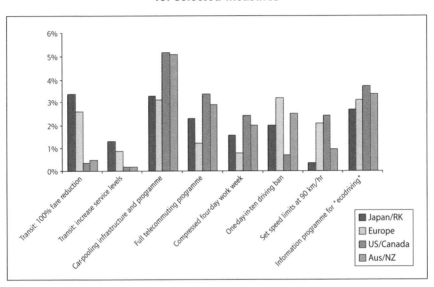

One example of the difference in the flexibility of the current systems is in the level of public transit infrastructure. IEA Europe and Japan/Republic of Korea (RK) tend to have greater levels of public transit and lower car ownership levels compared to North America and Australia/New Zealand (NZ). As a result, the measures to increase transit ridership result in significantly larger percentage reductions in petroleum use in Europe and Japan/RK relative to the other two regions.

On the other hand, car-pooling policies appear less effective in Europe and most effective in North America and Australia/NZ, where levels of solo driving are relatively higher (allowing a greater benefit from increased car-pooling).

The potential of telecommuting and flexible work policies also is least effective in the European region, relative to other regions. This is due to relatively lower current levels of solo car driving for commute trips. Thus, the benefit of a telecommuting or flexible work schedule policy is relatively greater in those countries that currently have more solo car commute trips.

On the other hand, driving bans appear most effective in Europe and least effective in North America. This is a function of the relative levels of household car ownership in each region. Average car ownership per household is highest in North America, which means that households are more likely to have at least one car available on any given day that a driving ban is enforced (as these are usually set by licence plate number).

Speed limit reduction and enforcement policies appear most effective in Europe and North America, where there is relatively higher motorway usage (relative to Japan/RK and Australia/NZ) and (in the case of Europe) higher maximum speed limits, providing more benefit from a reduction. Europe's results would be even higher except that we assume that all heavy trucks already travel at 90 kph, in accordance with EU law. Another fuel economy-related measure, "ecodriving" (campaigns to promote more efficient driving styles and vehicle maintenance), is assumed to have very similar levels of effectiveness across regions.

Implementation costs and cost-effectiveness

The costs associated with implementing each measure, and its cost-effectiveness on this basis, were also estimated. These are summarised in Table E-2, shown as an average cost per barrel of oil saved across the IEA and grouped in order of decreasing cost-effectiveness. (Separate cost estimates were also made by region and are shown in Chapter 3.) It is important to be clear that these results are based on relatively simple assumptions and are incomplete: they include only the direct costs incurred (mostly by governments) to plan for and carry out emergency measures. They do not include most costs or savings to travellers, such as for taking public transit or buying fuel (though fuel costs are included implicitly, since the costs are presented per barrel saved; therefore large fuel savings to consumers are the basis for a low cost-per-barrel estimate). The estimates also do not include many difficult-to-measure but important indirect costs and benefits, such as reduced or enhanced mobility, impacts on travel time (*e.g.* increases in travel time from lower speed limits) and safety (*e.g.* reductions in accidents and fatalities from reductions in speed limits). However, for those measures likely to have a significant impact in one of these areas, this is noted in the third

Table E-2

Summary of implementation cost-effectiveness of various measures

Implementation cost-effectiveness	Measure	Other potential impacts	Oil savings (from Table E-1)
VERY LOW COST Less than $1 per barrel saved	**Car-pooling:** large programme to designate emergency car-pool lanes along all motorways, designate park-and-ride lots, inform public and match riders		Very Large
	Driving ban: odd/even licence plate scheme. Provide police enforcement, appropriate information and signage	Possibly high societal costs from restricted travel	Very Large
	Telecommuting: large programme, including active participation of businesses, public information on benefits of telecommuting, minor investments in needed infrastructure to facilitate	Possible productivity impacts from changes in work patterns	Large
	Compressed work week (fewer but longer workdays): programme with employer participation and public information campaign		Large
	"Ecodriving" (efficient driving styles and vehicle maintenance steps): intensive public information programme	Likely safety benefits	Large
	Car-pooling: small programme to inform public, match riders		Moderate
LOW COST Less than $15 per barrel saved	**Speed limits:** reduce highway speed limits to 90 kph. Provide police enforcement or speed cameras, appropriate information and signage	Safety benefits but time costs	Large
	Driving ban: 1 in 10 days based on licence plate, with police enforcement and signage	Possibly high societal costs from restricted travel	Large
MODERATE COST Less than $50 per barrel saved	**Bus priority:** convert all existing car-pool and bus lanes to 24-hour bus priority usage and convert other lanes to bus-only lanes		Small
HIGH COST More than $100 per bbl saved*	**Telecommuting:** Large programme with purchase of computers for 50% of participants	Possible productivity impacts from changes in work patterns	Large
	Transit: free public transit (set fares to zero); 50% fare reduction, similar cost		Moderate
	Transit: increase weekend and off-peak transit service and increase peak service frequency by 10%		Moderate

* Note: no measures are estimated to cost between $50 and $100 per barrel saved.

column of Table E-2. Measures that are likely to have large indirect costs from restrictions on mobility are also likely to be relatively unpopular, making them more difficult to implement.

Another cost that is hard to measure is macroeconomic in nature. If application of one or more of these measures successfully reduces world oil demand sufficiently to result in a reduction in oil prices, this will yield macroeconomic benefits (or help avoid macroeconomic costs). Such impacts are very difficult to measure, but potentially quite large.

Thus, the estimation of costs associated with different measures is quite complex and is a subject that deserves a more detailed treatment than could be provided in this study. The cost estimates presented here may be most relevant for governments to understand how much various measures will cost them to implement. Even then, the specific costs of implementing a measure may be highly variable and subject to specific conditions and assumptions, and governments are urged to undertake their own detailed cost assessments.

The implementation cost-effectiveness of these assessed measures depends upon many factors, especially the amount of upfront investment made to implement them. In general, those policies that require significant investments or financial outlays are not likely to be cost-effective (roughly defined here as above $50 per barrel of petroleum saved, though there are none between $50 and $100). Those policies that are not cost-effective include decreasing public transit fares, increasing public transit service frequency, constructing car-pool lanes and purchasing home computers for half of all telecommuters. All of these involve substantial costs and their cost-effectiveness (i.e. more than $100 per barrel of oil saved) is likely to exceed any expected increase in the cost of oil during an emergency situation.

Those policies that are most cost-effective, with implementation costs less than $50 per barrel saved – and some much less – include information programmes to promote telecommuting and flexible work schedules, "ecodriving", car-pooling, odd/even day driving bans, and in some cases, speed reduction policies. Restriping of existing roadway lanes to create car-pool-only or bus-only lanes is moderately cost-effective, but significantly higher-cost than most of the policies focused on information campaigns. Odd/even day driving bans appear particularly cost-effective over a short period, despite costs associated with enforcing the bans. However, driving bans in particular may impose large indirect costs in terms of lost mobility. As mentioned, such losses are difficult to measure and no attempt has been made to do so here. In contrast, measures that provide more and/or better mobility options clearly provide benefits in this regard.

Conclusions and recommendations

There are a variety of potential policies and measures available to rapidly reduce oil demand in the transport sector. Some of these may make sense at any time, but might be easier to implement, be more effective, or be more cost-effective during an emergency situation such as an oil supply disruption. Though the effectiveness of each measure during an emergency is fairly uncertain and dependent on local circumstances, the available evidence suggests that some have the potential to significantly cut oil demand at a modest implementation cost. Savings on the order of one million barrels per day or more, on an IEA-wide basis, appear possible from well-conducted demand restraint programmes. This is enough to offset a fairly large reduction in world oil supplies.

This study represents one of the few recent, comprehensive efforts to identify and evaluate rapid "demand restraint" measures for transport. More work is needed to continue to improve our understanding in this area. Perhaps most important is for countries to conduct their own analyses, reflecting their own priorities and their national and local circumstances. This study provides methodologies and data that will hopefully be useful in that context.

Even lacking a precise understanding of all the issues related to this topic, it is important that IEA members and other countries have in place plans to respond to episodes of oil supply disruption, in much the same way as many now have systems for responding to periods of particularly bad air pollution. It is important to develop a careful, detailed plan, with public awareness and participation in order to help ensure that citizens will understand and accept the measures when actually implemented. It is also important that those measures with actions that must be taken in advance, in order to prepare for a possible emergency, are identified and the necessary pre-planning undertaken. For nearly every measure assessed in this report, some types of pre-planning and investments are required, without which the measure will likely be much less effective when actually implemented during an emergency.

Finally, when emergency episodes occur in the future, governments should carefully monitor their efforts and assess the effectiveness of their programmes, and share this information so that countries around the world continue to improve their approach and handling of such situations.

1. INTRODUCTION

This book aims to provide a better understanding of potential short-term oil demand restraint measures in the transportation sector, allowing IEA member countries and non-member countries alike to better prepare for unexpected oil supply constraints and price spikes. It identifies measures that appear likely to be effective, and cost-effective, and develops straightforward methodologies for estimating their effectiveness that can be used by countries to make their own assessments. It uses these methodologies to provide regional and IEA-wide estimates of the potential reductions achievable from various demand restraint policies, and makes it easier for countries to make their own estimates.

Why would countries act to restrain oil demand? The main reasons involve avoiding major disruptions in economic activities due to oil supply shortages and ensuring that existing supplies are allocated to the highest-value uses (by targeting demand restraint at lower-value uses). Even in the current era where oil prices react rapidly to changes in supply and demand, sudden, large supply disruptions could cause physical shortages for at least short periods of time. Further, the demand for oil is known to be highly inelastic in the short run – *i.e.* consumers and businesses do not react very quickly to changes in oil prices. Measures that help them to react faster, especially if such measures are low-cost, can help to reduce the economic impacts of disruptions and price spikes.

The types of measures appropriate for rapidly cutting oil demand in an emergency situation may have very different effects on travel and fuel consumption behaviour than would occur under normal circumstances. There may also be a greater variety of policies that are viable under emergency conditions than under normal circumstances, especially if they are applied in a temporary fashion. The travel demand literature, however, focuses mainly on estimating transport policy effects under normal circumstances. For example, promoting car-pooling under normal circumstances may achieve at best a modest effect due to poor response rates by commuters and other travellers, while under emergency conditions the response could be more substantial. This might occur for two reasons. First, some individuals may no longer have access to fuel or would face a long queue to obtain it and thus would actively seek out car-pooling options. Secondly, altruistic behaviour may be more likely during an emergency. If governments can assist drivers in their efforts to car-pool in these situations, it may simply help them to take actions they are interested in taking. For this reason, many of the estimates presented here, based mainly on historical data not relating to emergency situations, may underestimate the effectiveness of various policies in times of emergency. Ranges of estimates are generally provided, including the maximum potential savings that might be available. Consensus estimates of the most likely effect, based on our own judgments, are also provided.

Careful advanced planning appears critical to enable transportation demand management initiatives to be rapidly put in place. Continuing with the car-pooling example, one way of promoting increased car-pooling is to provide car-poolers with a travel time benefit, for example by providing special car-pool lanes on roadways. This requires significant pre-planning and some investments, in terms of preparing signage and lane markings in advance of any emergency, to indicate that during an emergency, such lanes could become car-pool-only lanes. This would allow rapid "deployment" of the lanes during an emergency, in combination with an enforcement strategy. These types of pre-planning activities are also discussed in the following chapters. Estimates of the relative costs and benefits of the various policies analysed include the costs associated with pre-planning and deployment during an emergency.

While many studies have examined travel behaviour under normal circumstances, few have focused on behaviour and how it may change during emergency conditions. The fuel supply emergency in the United Kingdom during 2000 serves as one example. We review some of the evidence on potential travel behaviour change based upon studies undertaken after the emergency. We also have sought to examine some of the effects of the global "energy crises" that occurred during the 1970s. Evidence from previous crises provides some basis for understanding both the potential of major behavioural changes to occur and to some extent the conditions that allow people to reduce their dependence on private car usage. In addition, we review some other contingency emergency fuel reduction plans to fully understand the planning measures that others have promoted and that some nations have adopted.

One key set of policies is treated somewhat differently from the others: those associated with the price of fuel. We assume that in most countries, if severe

Demand response measures: an economic perspective

The main benefit to most of the policies analysed in this report is increasing the flexibility of choice among travellers to respond to oil supply disruptions and/or price spikes. This can also be characterised as increasing the price elasticity of demand for transport fuel. This is shown graphically in Figure 1-1. The initial quantity of fuel demanded is Q1. Under an inelastic demand response, this would drop to Qi and with greater elasticity this would drop to Qe. Corresponding price effects are Pi for inelastic demand, which is greater than Pe when demand is more elastic. The economic consequences are best measured by changes in consumer surplus. For the more elastic case, the reduction in consumer surplus is the area with darker shading. For the more inelastic case, the reduction is this darker area plus the lighter shaded area. Thus, there is smaller reduction in consumer surplus and societal welfare when the elasticity of demand is larger. This should therefore be a primary goal of demand restraint measures – to increase the demand responsiveness of the transport sector to fuel price increases and/or supply constraints. However, the cost of the measure should be less than the benefit it provides in terms of reducing the loss of consumer surplus.

Figure 1-1

Effects of increasing elasticity of demand response

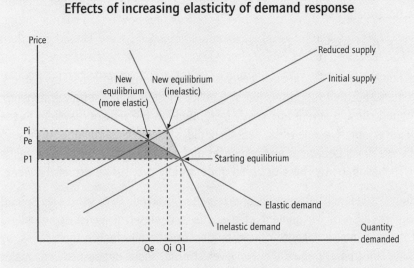

transportation fuel supply constraints occur, then prices will increase through market forces. Using policies to increase retail prices further may be counter-productive. (On the other hand, lowering retail prices, such as by quickly cutting taxes, may also be counter-productive as it dampens this price signal and could spawn shortages.) We review recent estimates of consumer response to price changes, to try to understand what the response may be to a price spike, with no government intervention. We note that fuel price increases can have different effects in different countries.

Previous fuel crises: what can be learned?

Short-term transportation fuel supply shortages are not an unusual occurrence. These have occurred several times in the past few decades due to political disturbances in the Middle East. More recently, refinery supply constraints (*e.g.* in California) and strikes (*e.g.* in the UK) have been blamed for shortages. Under normal market conditions, a shortage in supply would naturally lead to higher prices at the pumps, given no immediate reduction in the demand for oil. More commonly, various disruptions have occurred that offer an opportunity to observe changes in the behaviour of car users. We briefly review some of the evidence on how consumers react to these short-term disruptions. This provides some understanding of the potential for travel demand policies to help mitigate the impact of disruptions, and what happens without them.

The British fuel protests of 2000

In September 2000, a one-week blockade of British refineries by haulers, farmers and their supporters led to a major short-term fuel supply emergency. Though causing severe and costly disruptions to the transport system in the UK over about a one-week period, it did also serve as an opportunity to observe driver behaviour and responses under conditions of severe constraints on fuel availability. These occurred quite rapidly over the course of the week.

Many interesting behavioural effects were observed. First, shortages spread both because of curtailment of gasoline and diesel deliveries to refuelling stations, but also because drivers tended to stockpile fuel in their tanks, by filling up more frequently. However, traffic soon diminished on major motorways as people reduced the number of trips or their length. Eves *et al.*

(2002) evaluated traffic count data for several motorways and found significant drops during peak and off-peak times. For example, the M25 London Orbital Motorway experienced about a 23% drop in traffic during the morning peak and a 44% reduction during off-peak periods. This clearly suggests that those trips that are more discretionary (such as non-work trips) tended to be avoided relative to less discretionary peak hour trips. They also estimated that reductions in heavy-duty vehicles were larger than for cars on the M25, but this result was not consistent for the other motorways measured in their study.

Eves *et al.* also estimated changes in speed during the emergency compared to before the emergency. They found that during peak periods, speeds actually increased since there was less traffic congestion. However, during night-time periods they found speeds were marginally lower. They also found that overall average speeds were lower, after the reduced congestion effect was controlled for. This suggests that drivers may have tried to conserve fuel by reducing speeds.

Chatterjee and Lyons (2002) conducted a fast-response survey immediately following the fuel emergency to examine how behaviour changed. While their survey sample was non-representative of Great Britain as a whole, they did find some suggestive results. The main response of most people was to reduce the number of trips taken. The vast majority of these were classified as "other" trips, that is, they were not commute, business, school or grocery shopping trips. Commute trip reductions did occur and there was an increase in car-pooling for commute trips, as well as some shifting to other modes. School trips saw a major increase in walking. Overall, the main response seemed to be associated with reducing "other" trips which would tend to be more discretionary (and potentially lower-value) in nature.

A telephone survey that was conducted about two months after the fuel emergency analysed travellers' behaviour during the emergency (Thorpe *et al.*, 2002). About 29% of respondents reported actually running out of fuel during the one-week emergency. Most people (73%) continued to drive and there was about a 24% shift away from driving alone. Other modes also saw large shifts, including a 37% shift to walking and a 42% shift to public transit. There was a large percentage increase in the number of people who reported telecommuting (which was only three people before the emergency) rising to 19 during the emergency, out of the sample of 1 001 individuals.

Analysis of these same data by Noland *et al.* (2003) focused on how disruptive people thought a future fuel emergency would be to their engagement in activities. One of the interesting conclusions was that the vast majority of respondents did not expect a large amount of disruption, although key sub-groups did, especially for work-related travel. There was particular concern regarding maintaining fuel supplies for critical services, such as emergency services, providing supplies to hospitals, etc. But it appears that the diverse and widely available public transit systems in the United Kingdom are likely one reason why so many respondents felt that they could still engage in many travel-related activities even if there were severe fuel shortages.

One of the important conclusions that can be reached from observing the British fuel emergency is that, while many people were affected, most found ways to cope with travel needs over the duration of the emergency. Clearly the economic (and political) costs of supply disruptions and potential food shortages would have been too severe, indicating the dependence of society on transport and therefore on reliable fuel supplies. But perhaps the costs associated with the transport disruptions could have been lowered with a more systematic programme providing better travel and non-travel alternatives to the public during the emergency.

IEA member countries are under obligation to hold oil reserves of at least 90 days of the previous year's net imports (see box). Therefore, most externally-generated supply crises will allow some time for preparatory action. The British emergency, however, demonstrates that a diverse transport system and a diversity of integrated land uses can provide individuals with feasible options. It also demonstrates the need to prioritise how the fuel is distributed if supplies are critically short, which might require some sort of government-controlled allocation scheme to maintain basic economic necessities such as food deliveries. Apart from allocation actions, governments may want to consider measures for reducing travel demand – the main focus of this book.

Australian industrial dispute of 1981

In September 1981, an industrial dispute over the shipment of petroleum products led to the closure of the only oil refinery in South Australia. Purchasing restrictions were rapidly introduced, mainly to prevent hoarding of supplies. These included price (expenditure) limits on how much could be

Demand restraint as an emergency response measure

Emergency response is a main element of the IEA's treaty, the Agreement on an International Energy Program *(IEP Agreement). It includes the important commitment by IEA participating countries to hold oil stocks equivalent to at least 90 days of net oil imports. The IEP Agreement also defines an integrated set of emergency response measures, including "stockdraw" (use of emergency oil reserves), demand restraint, fuel switching, surge oil production and sharing of available supplies, for major international oil disruptions which reach the 7% threshold (the "trigger") defined in the IEP Agreement.*

In accordance with obligations laid out in the IEP Agreement, each participating country maintains at all times an effective demand restraint programme which can be implemented promptly in an emergency. This includes measures in transport as well as other oil-consuming sectors. In the event of an activation of IEP emergency response measures, each IEA member country will be expected to immediately implement demand restraint measures sufficient to reduce oil consumption by 7% of normal demand levels. In a more severe disruption, this could be raised to 10%.

Demand restraint measures are not exclusively reserved for disruptions which trigger an IEP response. For disruptions below this level, the IEA has a complementary set of measures known as Co-ordinated Emergency Response Measures (CERM). These provide a rapid and flexible system of response to actual or imminent oil supply disruptions. Under a collective action of the CERM, member countries would be expected to contribute either with the use of emergency stocks or other possible emergency response measures such as demand restraint.

In the context of the IEP, demand restraint refers to short-term oil savings which can be achieved during the period of an emergency. As emphasised throughout this volume, this should not be confused with energy conservation or medium- to long-term measures to reduce oil consumption.

Measures to achieve demand restraint fall into three main classes – persuasion and public information, administrative and compulsory measures and, finally, allocation and rationing schemes. The initial emphasis is likely to be on persuasion and light-handed end-use demand restraint measures rather than on compulsory measures or allocation. Some member countries may prefer, especially in the early phase of an emergency, to draw stocks in excess of their 90-day IEA commitment rather than introduce demand restraint measures, as allowed for in the IEP Agreement.

purchased, odd/even licence plate sale days and bans on refilling portable containers. This was followed by a weekend ban on sales and a coupon-based rationing system in Adelaide which lasted for four days. This was geared mainly to allow essential economic activities to continue; thus, coupons were provided mainly to truckers and other business vehicles, and private motorists were generally excluded. While details are unavailable, the rapid introduction of rationing clearly indicated that a contingency plan was in effect and this helped limit hoarding, which certainly would have worsened the economic damage from the emergency. It is unknown, however, how long the bans on sales to private motorists could have continued under longer-term emergency conditions without severe consequences (Lee, 1983).

The 1970s fuel crises

The 1970s global fuel crises (often called the "energy emergency") were clearly an important period for understanding how to deal with supply disruptions. Many of the transport demand management (TDM; also known as mobility management) policies still in use today were devised during the 1970s, although now these are normally justified for traffic congestion management or pollutant emissions reduction rather than for fuel conservation.

There were actually two main periods of fuel shortage in IEA countries – 1973-1974 and 1979-1980. The 1973 emergency was a result of the OPEC cartel cutting off supplies to Europe and North America, in response to the Egyptian-Israeli war occurring at that time. Oil prices quickly quadrupled (from $3 to $12 per barrel) and shortages ensued despite the price increase (the supply cut-off was nearly complete in countries like the United States and the Netherlands). The embargo ended after about six months, in March 1974, but prices continued to rise throughout the subsequent few years.

The second major oil emergency occurred in 1979, triggered by a revolution in Iran. Iranian oil output and exports dropped precipitously and quickly caused a significant shortage of oil around the world, with a sharp rise in world oil prices. The Iran-Iraq war caused a severe drop in Iraqi output in 1980, exacerbating the situation. Oil prices rose from $14 per barrel in 1978 to $35 in 1981 (in nominal dollars). Price controls in countries like the United States were lifted over this period, resulting in much higher retail prices but also eliminating queues for gasoline by 1981.

As mentioned, an important feature of the 1970s crises was that at that time many IEA countries had price controls on the sale of gasoline (Lee, 1983). The resulting inability of market forces to respond to a supply shortage, through increased prices, led to some "artificial" shortages as refiners in some countries exported gasoline to countries where prices were allowed to be higher. These types of price controls are now mostly gone, although it is certainly possible that some governments would reintroduce them should large price spikes occur. Pressure is often put on the government to lower taxes under periods of high underlying fuel cost, and this could certainly occur during a price spike episode. Such a lowering of taxes could have similar consequences as price controls, at least in terms of triggering physical shortages.

Hartgen and Neveu (1980) provided an assessment of transportation conservation measures undertaken in New York State during the 1979 emergency. They reported that New York reduced gasoline consumption, in aggregate, by 6% during the emergency, mostly through reductions in car driving. Drivers found other ways to move around: increases in public transit usage accounted for 31% of total fuel use reductions in New York City, while in the rest of the state, the combination of public transit and ride-sharing accounted for 24% of the savings. Switching to use of a more fuel-efficient car was also important (such as through shifting by multi-car households from one car to another). The study emphasised that these savings were achieved predominantly through voluntary action, and that significant additional conservation could be achieved through government programmes.

This review of the literature uncovered few additional available studies examining how driver behaviour responded during the 1970 crises. Hartgen and Neveu (1980) reference various government-funded reports, but most of these are out of print.

Lessons learned

Clearly, one of the key lessons of the previous crises is that travel demand can be reduced (or drop on its own accord) fairly quickly during a supply emergency. The British emergency in particular showed some remarkable short-term effects, especially with regard to non-essential trips. However, the ability to sustain these sorts of reductions over a longer term of several weeks or a few months could be far more difficult, at least without high economic costs. The evidence from the 1970s suggests that some moderate reductions were achievable without any government action other than appeals to altruistic behaviour. Clearly, in cases where fuel simply was in short supply, reductions in travel were forced and it is unclear at what cost to society. Thus a key question is, can governments intervene in ways that make it easier and less costly for travellers to cope with supply disruptions and price spikes? Subsequent chapters of this book look at this question.

Previous emergency planning efforts

Member countries of the IEA, through their "National Emergency Sharing Organisations" are responsible for implementing emergency measures. Member countries are required to have oil demand restraint programmes in place that can reduce oil demand by 7-10% in the event of a supply disruption. Most oil demand is now generated by the transport sector. Therefore, any strategy to reduce oil demand must involve emergency planning to reduce transport oil consumption. While previous emergency planning efforts in the transport sector have been carried out, many of these have not been updated in many years. Some of these are reviewed below.

In the United States, following the 1973-1974 energy emergency, there was a flurry of activity among urban areas to develop "Energy Contingency Plans". Much of this planning effort was focused on gasoline supplies and availability, but some also covered heating oil supplies. For the most part, the transport-related measures focused on many of the transportation demand management (TDM) measures discussed in the following sections of this report. These included car and van-pooling programmes, flexible/compressed work weeks and various ride-sharing programmes. Other actions were related to gasoline sales, such as odd/even licence plate purchase days (Barker, 1983).

Various public transit-related actions were also planned for. These were intended to put more buses into service by activating reserve fleets, using school buses and changing maintenance schedules. The dissemination of information on public transit was also planned for (Barker, 1983).

Barker (1983) reports that many of these measures were considered unsuccessful. The exception was efforts to control queuing at gasoline stations by odd/even purchase days, which, while effective at reducing queues, only had minor effects on total fuel consumption. Car-pooling programmes suffered from the time taken to set up the technology to match riders and contact people. This is an important observation. With modern computer technology and experience gained (at least in the United States) at running car-pooling programmes, it is likely these could be set up much more quickly – and many already exist. In any country, setting up a system to help match travellers with others, to help car-pools form quickly, could yield important benefits during a supply emergency. Developing an infrastructure of high-occupancy vehicle (HOV) lanes – that is, lanes that become dedicated for car-poolers during a supply emergency – could also facilitate greater success at instituting a short-term car-pooling programme. The potential impacts of such measures are discussed in the following chapters.

Several difficulties were reported with changing public transit operations on short notice. These included the inability to quickly train new drivers, insurance-related problems with using school buses for public transit, and the lack of sufficient reserve vehicle capacity to bring on line quickly. If these sorts of measures are to be implemented, planning and actual investments need to be made well in advance of the emergency occurring. Provision of public transit information was hampered by not having enough telephone capacity to handle calls. Internet-related technologies could make the dissemination of this type of information far more efficient today.

Lee (1983) provides some perspective on IEA planning efforts in response to the 1970s crises, based largely on personal communications and unpublished reports. One issue is that in the 1970s, many IEA member countries were more interventionist with various price controls on gasoline, government ownership of refineries and much greater political support for interventionist policies. At least in Europe, the European Union now generally discourages or prohibits policies aimed at controlling prices or nationalisation of industries.

Norway has had detailed contingency plans going back to the 1970s. Much of their planning relied on voluntary conservation measures, but if an emergency worsened over time, weekend driving bans, rationing and curtailment of recreational travel were planned for (Lee, 1983). Many countries had contingency plans to introduce rationing if supplies became constrained, including having stockpiles of ration coupons already printed.

In 1979, New Zealand implemented a carless day scheme, based on coloured stickers put on vehicles. This banned car use one day a week. Service stations were also closed on weekends. In 1981 the parliament passed the Petroleum Demand-Restraint Act which authorised both of the above measures during emergencies as well as odd/even day sales, maximum and minimum purchase limits and the printing of ration coupons.

More recently, ICF Consulting developed a draft emergency plan for Greece (see box). Various transport-related demand measures were included in this plan to be implemented in the event of an oil supply emergency. The Emergency Plan calls for a series of actions to be undertaken by the central government, grouped in sets dependent on the severity of the emergency. The transport sector components of this Emergency Plan are outlined in the box.

One feature of this plan is that initial actions are relatively minor, consisting only of public relations efforts and calls to conserve energy. Only if this proves ineffective would more severe actions to mandate reductions in private car use be implemented. Clearly, if the price of fuel is high, this will in itself result in some reductions in consumption. On the other hand, if a supply emergency were multi-national in its scope, reductions in Greek demand for fuel might have little impact on the underlying oil price, in which case even the most severe cutbacks in consumption might not yield price reductions.

Key components not considered in earlier planning efforts were the institutional and management requirements of actually implementing a plan (Barker, 1983). In other words, it is one thing to mandate that more commuters car-pool and quite another thing to actually have a system in place to ensure that this happens. A key component of any emergency plan should be to lay the foundation such that measures can be quickly and successfully implemented during an actual emergency. The experience in implementing travel demand restraint and mode-switching policies for other goals (such as traffic congestion reduction and air quality improvement) helps provide such a basis.

Various planning elements are necessary to implement the many policies discussed here. While some can be implemented quite quickly or will occur naturally due to price increases, others can be best facilitated by comprehensive pre-planning that builds more flexibility into the transport

Energy emergency plan for Greece

Greece's energy emergency plan has three sets of measures. At the lowest level of emergency, the initial implementation of the plan focuses on voluntary actions that are believed to be of low cost and cause few, if any, distortions in the market:

1. **Voluntary demand reductions.** The Minister of Development authorises a public campaign on radio, television, newspapers and any other appropriate medium to call on travellers to volunteer to:
   a. Take public transitation and use car-pooling to the extent possible.
   b. Reduce the amount of driving, make efficient trips and walk instead of driving on shorter trips.

If further steps are required, the Minister may order the following actions that are considered more costly and could result in distortions to the economy.

2. **Mandated demand restraints.** Demand restraints considered will include restricting private automobile use, restricting service station operations and restricting the operation of energy-intensive industries.

A continued shortage or steady rise in prices will be a signal for further more severe actions to be taken.

3. **Additional mandated demand reductions.** After considering the effectiveness of the cumulative actions and the impact on prices, the Minister of Development can mandate demand reductions via:
   a. Reductions in speed limits.
   b. Further restrictions on private vehicle transportation, especially automobiles.

In all cases, the Minister will delegate monitoring and enforcement authority over all road transportation restrictions to the national police. Failure to comply with regulations will result in various fines and penalties.

system. Examples include premarking of motorway lanes for car-pools (with associated signage that indicates that such lanes would become car-pool-only lanes under certain circumstances), increasing the size of public transit fleets with reserve buses and bus drivers, obtaining commitments from employers that they will institute flexible work schedules and telecommuting plans, installation of variable speed limit signs for motorway and other high-speed road systems, and preparation of complementary publicity measures to inform people of the system and convince them of the benefits of reducing fuel consumption and how their behaviour is important.

2. ANALYSIS OF POTENTIAL POLICIES AND MEASURES

This chapter describes and analyses various transport demand restraint policies that could be applied under emergency conditions. While specific policies that could be implemented are fairly numerous, we focus our discussion on those that appear to be the most promising, cost-effective and politically feasible, distinguishing between several general categories of policy. Each set of policies is then analysed to determine the potential range of effectiveness. Effectiveness estimates are made for each IEA region using the best available data. In cases where data are not available, or may not be available for a specific region, we use the best assumptions possible. These are clearly stated in all cases. The estimates presented here are intended to be indicative – to provide order-of-magnitude indications – and to provide policy-makers with methodologies and guidelines for developing and scoring policies that may be appropriate in their country's context.

Transport demand restraint policies: overview and methodologies

"Transport demand restraint" policies are generally similar to "transport demand management" (or TDM) policies, a more common term in the transport literature, though these are usually thought of and applied not for emergency situations, but for managing transport demand and fuel use under normal circumstances, over time. The vast majority of the literature on the impacts of TDM policies relates to this more general situation, rather than emergency conditions. The impacts of TDM policies may be different in emergency than in normal circumstances, but it is still useful to gain an understanding of the types of impacts they have in the more general case. We first present some general results for various policies and then discuss and analyse the key policies in more detail, in the context of a fuel supply emergency.

Meyer (1999) reports results from an Apogee/NARC study done in the early 1990s that reviewed various estimates of the effectiveness of TDM policies.

This study provides estimates of percentage reductions in vehicle miles of travel resulting from various policies. TDM policies that involve increasing the price of transport were, not surprisingly, found to be more effective than policies that simply provide increased choice of travel options. Table 2-1 displays the range of estimated effects for selected TDM options.

Details on how these estimates were made and what specific conditions they refer to were not available. These were essentially based on a literature review conducted in the early 1990s. They do serve, however, to show reasonable ranges of effects on vehicle miles of travel (and consequently fuel consumption). Some of these policies are discussed further below.

A study by the German Institute for Economic Research (DIW, 1996) analysed the effectiveness of various measures that discourage or forbid vehicle usage. These are somewhat different from typical TDM policies which are usually aimed at providing travellers with additional choices or implementing market-based pricing mechanisms, but might be quite effective during emergencies. Estimates from the DIW study are shown in Table 2-2. These are based primarily on assumptions, rather than empirical estimates. However, they serve to highlight another set of alternative policies for reducing vehicle fuel consumption.

Table 2-1

Estimated effects of transport demand management

Measure	Percentage reduction in daily VMT	
	Minimum	Maximum
Employer trip reduction	0.2	3.3
Area-wide ride-sharing	0.1	2.0
Public transit improvements	0.1	2.6
HOV lanes	0.2	1.4
Park and ride lots	0.1	0.5
Bike and walk facilities	0.02	0.03
Parking pricing at work	0.5	4.0
Parking pricing: non-work	3.1	4.2
Congestion pricing	0.2	5.7
Compressed work week	0.03	0.6
Telecommuting	-	3.4
Land use planning	0.1	5.4
Smog/VMT tax	0.2	0.6

Source: Apogee/NARC study as reported by Meyer (1999); VMT: vehicle-miles travelled.

Table 2-2

Potential fuel savings from transport demand management

Measure	Percentage of total domestic fuel sales		
	Gasoline	Diesel	Total
Public appeals to reduce consumption without price effects	1.9	0.2	1.1
Public appeals to reduce consumption with price effects	7.6	0.7	4.6
Ban on motor sport events	0.1	0.0	0.0
Ban on driving by car to large-scale events	2.5	0.6	1.7
Speed restrictions[1]	7.2	1.7	4.8
Ban on driving every second Sunday	3.7	0.9	2.5
Ban on driving every second weekend	4.8	1.1	3.2
General ban on Sunday driving	9.3	2.2	6.3
Restriction on use by administrative decree[2]	5.5	1.3	3.7
Restriction on use by registration number[3]	3.6	0.9	2.4
General ban on weekend driving	12.6	3.0	8.5
Implementation of fuel supply ordinance (rationing)[4]	12.6	3.0	8.5

[1] 100 km/h on motorways, 80 km/h on other roads outside built-up areas.
[2] Public authorities set days on which drivers are banned.
[3] On each weekday two final registration numbers banned.
[4] Savings of 15% in journeys to work/training/education, of 7.5% in business travel and of 90% in shopping, leisure and holiday travel.
Sources: Branch Association of the Petroleum Industry, Germany; DIW (1999).

The general approach taken in the analyses that follow is based on estimating a range of possible effects for each of the potential policy approaches. The aim is not to specify a specific policy, but rather to examine a general strategy, such as "increasing car-pooling" or "reducing travel speeds" and then to examine the maximum potential fuel savings possible if this can be achieved.

Within these analyses we also consider how actual policies under conditions of a fuel emergency might lead to actual reductions. For example, a policy of designating various motorway lanes as car-pool lanes could lead to a reduction in fuel that is somewhat less than if all trips now had more than one occupant in the car. Therefore, in most cases, the maximum potential is unlikely to be achieved when put in the context of actual policies that can be implemented.

The other consideration, however, is that behavioural responsiveness to policies is likely to be more effective under emergency conditions. This is true for several reasons. First, altruism on the part of individuals is likely to be high, at least in the short term. Second, supply disruption-induced fuel price increases will provide an incentive to reduce fuel consumption for financial reasons. Finally, actual shortages would naturally force some people to respond to policy initiatives. Therefore, at least in some cases, our estimates may be conservative, as they are to a large degree based on behavioural responses to policies under normal non-emergency conditions.

Our analysis also seeks to use the best available data, as previously described. Where possible, data and estimates from individual countries or regions are used. Assumptions are clearly noted where data are not available. Details on many of the data sources used in this analysis are presented in the appendix.

Pricing and taxation-related policies

There is a large variety of pricing policies, ranging from fuel taxes that can have a direct impact on fuel consumption, to more esoteric measures with limited local impacts, such as congestion pricing, or various measures to increase the "opportunity cost" of parking.

A substantial literature exists on the price responsiveness of fuel consumption to changes in prices. This is known as price elasticity of demand and is defined simply as the percentage change in the amount of fuel consumed for a percentage change in the price of fuel. For example, a price elasticity of –0.3 means that a 10% increase in price results in a 3% decrease in consumption.

There have been several recent reviews of the literature on fuel price elasticities. These include Goodwin *et al.* (2004) and Graham and Glaister (2002). Both studies were funded by the UK Department for Transport and provided very similar assessments of the average estimates of fuel price elasticities in the literature. The consensus range is that short-run fuel price elasticities are between –0.2 to –0.3, with long-run elasticities being between –0.6 to –0.8. The distinction between long-run and short-run elasticities is somewhat ambiguous and is partly related to the estimation techniques used. From a time perspective, the short-run effects occur almost immediately, while the long-run effects occur in time scales related to the turnover of the vehicle fleet and relocation of activities within an urban area (probably about 5 years on average).

The behavioural effects associated with short-term elasticities are generally less driving, more efficient driving styles and more efficient allocation of trip-making decisions (for example trip chaining). The price elasticity literature does not tend to disaggregate these effects. Longer-run effects are associated primarily with purchase of more efficient vehicles and, to some extent, with relocation and redistribution of activities and land uses to shorten trips.

Changes in fuel prices also have an effect on total kilometres travelled. Consensus elasticity estimates for this effect are also found to range from –0.15 in the short run to –0.30 in the long run. The short-run effect is somewhat similar to the short-run fuel consumption effect. Interestingly, this effect is smaller and the difference could perhaps be attributed to changes in driving style that can also lead to fuel consumption reductions. If disaggregated in this way, we could say that the direct short-run effect from a price increase (due to less driving) is –0.15, while the effect from changes in driving style is between –0.05 and –0.15 (based on the difference in elasticities).

Another important consideration is how changes in travel time affect demand for car travel and indirectly fuel consumption. Noland and Lem (2002) reviewed the literature on how changes in road capacity affect total travel (essentially what is known as the induced demand effect). While not explicitly considering the travel time effect, the consensus estimates on induced travel elasticities (expressed as changes in vehicle-miles or kilometres travelled [VMT or VKT] with respect to changes in lane miles) is about 0.2 to 0.3 in the short run, ranging from 0.7 to 1.0 in the long run.

More explicitly, looking at travel time elasticities, Graham and Glaister (2005) report that these are about –0.20 in the short run and up to –0.74 in the long run. One implication that they highlight in their review is that increasing travel times and congestion will tend to be more important, in the long run, than increases in fuel prices, in offsetting vehicle-kilometres of travel (VKT) growth.

Graham and Glaister also reviewed elasticities of road freight demand. This is normally expressed as changes in tonne-km for a given change in generalised cost, of which fuel prices would be one component. They found wide variation between different commodity types and no easily identifiable average value. Their main conclusion is that the elasticity is negative and in some cases could be quite large, which contradicts assertions that freight demand is relatively inelastic with respect to price changes.

A recent European Commission project, TRACE, also estimated and reviewed travel demand elasticities. The basic approach taken by this project was to use national travel demand modelling systems from various countries. Within this context, the travel demand elasticities are dependent on many of the modelling assumptions made and should be considered in this light. However, they do provide more detail than aggregate econometric studies. That detail includes extensive elasticity estimates for different types of trips and also for parking pricing policies. These are reported in *The Elasticity Handbook* (TRACE, 1999).

Key results from the TRACE project are presented in Table 2-3. These give elasticity estimates for how VKT changes with changes in fuel price, travel times and parking charges. Parking elasticities include an average estimate based on increasing existing parking charges and new charges prorated to the distance travelled. Each is disaggregated by trip purpose. What is especially interesting about these results is the different responses for commuting and business trips versus "other" trips. This last category would include most trips which are more discretionary in nature. Clearly, these will tend to be affected much more by these type of charges, at least in the short run. Long-run responses are higher in all cases, especially for travel times.

Table 2-3

Key results from the TRACE project

Trip purpose	VKT with respect to		VKT with respect to parking charges				
	fuel price	travel time	average	distances 0-5 km	distances 5-30 km	distances 30-100 km	distances over 100 km
Short term:							
Commuting	-0.15	-0.48	-0.02	-0.10	-0.02	-0.01	-0.01
Business	-0.02	-0.05	0	0	0	0	0
Education	-0.06	-0.05	-0.01	-0.12	-0.02	0	-0.00
Other	-0.22	-0.19	-0.08	-0.30	-0.06	-0.01	-0.02
Total	**-0.15**	**-0.28**	**-0.03**	**-0.18**	**-0.03**	**-0.01**	**0**
Long term:							
Commuting	-0.25	-1.04	-0.04	-0.13	-0.06	-0.02	0
Business	-0.22	-0.15	-0.03	-0.02	-0.02	-0.03	-0.03
Education	-0.38	-0.84	-0.03	-0.17	-0.06	-0.01	0
Other	-0.47	-0.86	-0.16	-0.36	-0.18	-0.05	-0.00
Total	**-0.31**	**-0.80**	**-0.07**	**-0.22**	**-0.10**	**-0.03**	**-0.02**

VKT: vehicle-kilometres travelled.

Table 2-4 also presents modelling results on the effectiveness of some individual pricing policies (US EPA, 1998). These are based on cities in California in the early 1990s. Results show percentage reductions in vehicle-kilometres of travel (VKT), trips, travel time and fuel usage.

Table 2-4

Modelled estimates of pricing measure impacts

Policy	Percentage reductions			
	VKT	**Trips**	**Time**	**Fuel**
Region-wide congestion pricing	0.6-2.6	0.5-2.5	1.8-7.6	1.8-7.7
Region-wide employee parking charges				
$1.00 per day	0.8-1.1	1.0-1.2	1.0-1.1	1.1-1.2
$3.00 per day	2.3-2.9	2.6-3.1	2.5-3.0	2.6-3.0
Gasoline tax increase				
$0.50 per gallon	2.3-2.8	2.1-2.7	2.4-2.8	5.8-7.4
$2.00 per gallon	8.1-9.6	7.6-9.2	8.4-9.7	24.3-27.3
Mileage and emissions-based registration fees				
Fee range from $40-$400 annually	0.2-0.3	0.1-0.2	0.2-0.3	3.4-4.4
Fee range from $10-$1 000 annually	2.9-3.6	2.7-3.3	2.7-3.5	6.3-7.9
VMT fee of $0.02 per mile	4.6-5.6	4.4-5.4	4.8-5.7	4.8-5.7

Source: US EPA (1998); VKT: vehicle-kilometres travelled.

These results provide a basis for developing simple methods to evaluate short-run responses to policies that affect fuel prices, travel times and parking charges. In general, these effects will differ depending upon trip purpose. This study is focused on very short-term and rapid responses to fuel shortages. The elasticities reviewed here are all based upon econometric or modelling results which may define "short term" less explicitly. For example, econometric approaches generally assume that short-term elasticities are derived from cross-sectional studies or from lagged estimates that separate short and long-run elasticity coefficients. Sometimes, the actual time frame is ambiguous. In general, however, "short term" is anywhere from a few months up to a year, relative to "long term" which could be in the range of 1-10 years (or however long it takes to turn over the vehicle stock and for relocational effects to occur).

The short-term elasticities in this case should probably be viewed as lower bounds for the type of very short-term policies that might be considered in the context of short-term demand restraint measures (*i.e.* a few weeks of altered behaviour due to price increases). There is some evidence that effects can be significant when the crisis is limited in duration, as discussed in the context of the British fuel emergency (Noland *et al.*, 2003).

Fuel prices under emergency conditions and regional variation in responses

Assuming that countries allow market forces to operate in the petroleum sector, any reduction in supply should give rise to increases in the price of fuel. This of itself will tend to dampen demand and induce many of the behavioural changes sought by implementation of demand restraint measures. These include shifts in mode of travel, reduced trip-making and reductions in travel speed, amongst others. Therefore, to some extent it is not strictly necessary to implement pricing policies that increase fuel prices.

However, one of the key issues is that each IEA region will tend to have somewhat different responses to price increases based upon the variation in transport infrastructure, the ability to offer alternative modes of travel and the existing taxation schemes in each country.

Fuel taxation tends to vary both between countries and between IEA regions. In general, the United States has the lowest level of fuel taxes (and prices), while European countries have the highest tax levels and price levels. Canada has taxes that are about double those in the United States, but prices are only about 20-30% higher. Prices in Australia and New Zealand are similar to those in Canada, but the share of the price that is taxes is slightly larger. In Europe, prices can vary by as much as 30% between the low-price (low-tax) countries such as Greece, and those such as the United Kingdom and Denmark, with higher prices and taxes. Prices in Japan and the Republic of Korea are similar to average European values. Thus, in general, we can characterise North America as being low-price and low-tax, followed by Australia and New Zealand being slightly higher, and Europe and Japan/Republic of Korea being the regions with the most expensive fuel.

What this means in terms of demand restraint in each of the regions is that fuel price increases in Europe and Japan/Republic of Korea will be less effective by themselves in reducing overall demand compared to North

America and Australia/New Zealand. As mentioned, demand elasticity is the percentage change in demand in response to a percentage increase in price. Since most tax regimes have a fixed tax per litre, this means that for those areas with relatively high taxes, underlying fuel price increases will have a smaller percentage effect on retail fuel prices (unless the tax is set as an *ad valorem* tax and thus increases in proportion to fuel price). This difference far outweighs the likely differences in elasticities, resulting in a smaller effect in reducing the consumption of fuel. Some tax regimes include value-added or sales tax on the total, but generally this is smaller than the fixed tax rate.

The other complication is that those countries that have historically had higher taxes also tend to have developed less reliance on private car travel. This means that, in general, they will have more compact and mixed-use development, more public transit and lower levels of car ownership. In this sense, though the previous discussion suggests that the total price increase will be a lower percentage, they also tend to have more elastic demand (since there are more travel options).

Table 2-5

Effect of a 50% increase in fuel price on demand under different conditions

Tax percentage of retail fuel price	Change in retail fuel price from 50% increase in petroleum price	Fuel use elasticity and resulting percentage fuel use reduction			
		-0.1	**-0.2**	**-0.3**	**-0.4**
20%	40%	-4%	-8%	-12%	-16%
40%	30%	-3%	-6%	-9%	-12%
60%	20%	-2%	-4%	-6%	-8%
80%	10%	-1%	-2%	-3%	-4%

Table 2-5 shows some of the effects related to elasticity of fuel consumption with respect to price and the percentage of the total price that consists of fuel tax. As the first two columns show, as the percentage of the retail price composed of tax increases (with higher tax rates), the impact of a change in the underlying product price on final retail price diminishes (since the tax does not change if it is nominal, which is the case in most countries). The following columns show the percentage reduction in demand that occurs with different elasticities and different changes in final fuel price. As can be seen, when the amount of taxation is less, one gets a larger percentage reduction in consumption for a given elasticity. If fuel taxes consist of 20% of the total price (similar to United States values) and we assume a relatively inelastic response of –0.2, then the percentage reduction is –8% (fuel use is 4% lower than it would be without the tax). With a higher tax, such as one that represents 60% of the retail fuel price (similar to many European countries), then even with a higher elasticity, such as –0.3, the reduction in fuel use could be lower (in this case 6% rather than 8%).

Another unknown factor is how severe shortages could lead to exceptional increases in gasoline prices. Under these circumstances, constant elasticity conditions may no longer hold, as consumers may face real budget constraints (income effects) in purchasing fuel. This could imply far larger reductions, overriding any effects from existing tax policies.

While it is likely that price increases will have some effect in reducing consumption and equilibrating demand and supply, governments may be under pressure to reduce fuel tax levies. The British fuel protests of 2000 received their initial spark from spikes in the price of fuel, not any recent government policy with respect to taxes (although the fuel tax escalator had been pushing up fuel tax levies above the rate of inflation for several years). While prices came down eventually, the government made small changes in fuel taxes in response to the protests and also eliminated the automatic fuel tax escalator.

Government fuel tax policy should be careful not to offset price increases due to supply constraints, as this will only be counter-productive and could exacerbate any spot shortages of fuel. Since these price increases will tend to be automatic, the key policy lesson is that fuel taxes should not be used to offset price increases. There could clearly be political incentives for some governments to follow a counter-productive strategy such as this.

The other major point of this analysis is that initial higher fuel tax rates also tend to automatically mitigate the effects of increases in price. This is due both to the likely higher level of alternative transport infrastructure available, but also is related to the proportional increase in fuel prices (assuming elasticities are constant).

Implementation of pricing policies

Various road pricing policies can also be effective at reducing fuel consumption. Many of these cannot be implemented without sufficient pre-planning. Some of these are discussed below but no analyses of effects are provided. The various elasticity estimates provided above can be used by those wishing to estimate the potential effects of these types of policies.

Road pricing policies can be implemented in several ways. For example, one simple mechanism is a direct fee based on vehicle-kilometres travelled (VKT). This could be implemented by basing annual registration fees on VKT. Another method is through insurance premiums, sometimes called "pay as you drive" (PAYD) or "pay at the pump" (PATP) insurance schemes. These schemes have been estimated to reduce driving by shifting fixed costs to variable costs (Litman, 2000). Actual reductions could easily be estimated from the VKT elasticities presented above. While collection of fees during annual registration would not be amenable to short-term increases in price, PAYD could be if vehicle movements are tracked in real time, as many of these schemes have proposed. These types of schemes would make it feasible to institute surcharges for short periods of time in response to a need to reduce fuel consumption.

Congestion pricing, primarily aimed at reducing congestion, may also provide some reduction in fuel use. This depends on how the scheme is designed. For example, the London congestion charging scheme levies a £5.00 charge for vehicles entering Central London between 7:00 a.m. and 6:30 p.m. on weekdays. Recent estimates found that this scheme has reduced traffic in Central London by about 30%. Estimates suggest that about 50% of those previously driving to Central London have switched to public transit, about 15-25% have switched to cycling, motorcycling and car-pooling. Overall car occupancy is estimated to have increased by about 10%. Many trips that previously went through the zone have now been diverted around it. On the basis of these initial estimates, it is likely that vehicle travel and fuel consumption have decreased, although it would be difficult to give precise estimates (Transport for London, 2003). However, one important consideration is that when these types of systems are in place, it is relatively easy (at least technically) to vary the price under emergency conditions to further reduce vehicle travel for short periods of time. The scale of the London scheme is relatively small, so any net reductions in fuel consumption would also be small relative to total national consumption. But a system in place on the entirety of a nation's motorways might be able to deliver large reductions in an emergency.

The impact of parking pricing (or taxes) can also be evaluated from the elasticities above. Another parking policy is known as "parking cash-out". This is essentially a way of creating an opportunity cost associated with what is currently free employer-subsidised parking. This type of policy requires employers to offer all employees the cash equivalent of the value of the free parking that is offered, in lieu of that parking. This provides a strong incentive for employees to reduce the amount of driving for work trips (by forgoing parking and taking the cash). In a case study of eight firms that implemented this policy in California, Shoup (1997) found that vehicle-miles travelled (VMT) decreased by 11% with the share of solo commuter driving decreasing from 76% to 63% amongst the employees. Employees shifted to other modes, with car-pooling seeing the greatest increase in modal share.

Provision and promotion of alternative modes

One set of policies to reduce car usage is to encourage travellers to use alternative modes of travel. This includes shifting travel to public transit, car-pools, walking and bicycling. Policy mechanisms for accomplishing these types of shifts have been extensively explored over the last 30 years. One of the most effective means to encourage these mode shifts is to do so indirectly, by increasing the cost or decreasing the ease of car travel. These effects have already been discussed in the section on pricing policies, and are implicitly covered under the consequences of driving bans, speed reductions, etc. This section looks at other policies to directly increase the attractiveness of these modes, by making them less costly or easier for people to use, either by increasing the level of service or removing barriers to usage. This section discusses and analyses the potential of some of these policies.

The impact of public transit improvements on reducing car travel, which include a bundle of potential policies, can be quite difficult to estimate. Table 2-1 shows estimated vehicle travel reductions that range from 0.1% to 2.6% for a broad range of public transit promotion measures, which is a large range, though of relatively small magnitude. The details on what transit "improvements" this encompasses and the spatial scale of the travel reductions are not known. Improvements can consist of increases in scheduled frequency, spatial coverage, comfort, reduced crowding, improved information provision, as well as fare decreases.

Because of the wide range of potential effects, we have explored three main approaches most applicable for implementation on an emergency basis during a petroleum supply emergency. These three are fare reductions or elimination, off-peak service enhancements (service frequency increases) and bus lane prioritisation enhancements, discussed below.

One important conceptual issue that spans these three strategies is estimating their effect on private VKT. Typically and understandably, each has been assessed for its effectiveness in increasing public transit ridership. Although some studies take some or even all of the connecting steps, these are several steps removed from assessing the public transit passenger-km increases, private vehicle passenger-km decreases and private vehicle VKT decreases necessary to estimate petroleum demand reductions. Where available, we utilise studies that do estimate the private vehicle travel reductions directly (typically through the use of cross-price elasticities rather than just own-characteristic elasticities). In the other cases we must rely on assumptions, described below, to estimate these relationships.

Public transit fare reductions

The own-price demand elasticity of public transit patronage with respect to fare changes is well established, though based mainly on studies in North America. This elasticity is generally about –0.3, meaning that a price reduction of 10% yields a ridership increase of 3%.

Litman (2004) conducted a review of the literature and found that it breaks down to a –0.42 elasticity for off-peak travel and –0.23 for peak periods. According to a fact sheet from the Commission for Integrated Transport (2002b), in the United Kingdom since the 1990s, local bus fares have increased by 24% and local bus use declined by 11%, which would imply an elasticity of –0.46, though many other factors also changed during this time period. A study by Booz Allen Hamilton (2003) for the Department of Urban Services in Canberra, Australia estimated that, for bus users, own-price elasticities were –0.18 during peak and –0.22 during off-peak times. These findings show that commuter trips are less elastic than off-peak trips, *i.e.* that commuters are less responsive to price changes than riders at other times.

Nijkamp and Pepping (1998) report on a European analysis of public transit elasticities. Table 2-6 shows the results of their survey of four European countries. These elasticity values reflect changes in public transit trips and person-km. One of their conclusions is that the level of the elasticity varies by country, perhaps due to different situations in each country with respect to levels of urbanisation and availability of alternative modes (such as cycling in the Netherlands). Goodwin (1992) suggests that higher elasticities such as these may represent long-run rather than short-run effects (with short-run elasticities more appropriate for an emergency response). Dargay and Hanly (1999) similarly suggest a –0.2 to –0.3 short-run elasticity and a –0.4 to –1.0 long-run elasticity, with higher values for rural bus and intercity coach services. In any case, most of the European elasticities are higher than the –0.3 value commonly used in the United States. This suggests that Europeans may have more flexible travel options than Americans and are more likely to change modes if prices change.

Several studies of employer-paid commuter public transit benefits (the equivalent of reduced or free fares) have found substantial increases in public transit use through these programmes. As shown in Table 2-7, studies of the TransitChek programme in New York City and Philadelphia regions and of the

Table 2-6

Survey of public transit travel elasticities in four European countries

Country	Year of data	Competitive modes	Person-km elasticity	Trip elasticity
Finland	1988	2		–0.48
Finland	1995	3		–0.56
Finland	1966-1990	1	–0.75	
Netherlands	1984-1985	2		–0.35 to –0.40
Netherlands	1980-1986	2		–0.35 to –0.40
Netherlands	1950-1980	1	–0.51	
Netherlands	1965-1981	1	–0.53 to –0.80	
Netherlands	1986	2	–0.77	
Netherlands	1977-1991	2	–0.74	
Norway	1990-1991	3		–0.40
Norway	1991-1992	5		–0.63
United Kingdom	1991	4		–0.15

Source: Nijkamp and Pepping (1998).

Table 2-7

Change in public transit use due to employer-provided public transit benefits

Region	Type of trip	Percentage of employees reporting increased transit use	Average increase in weekly public transit trips per employee (employees receiving benefit)
San Francisco Bay Area	Commute	34%	2.1
	Non-work	29%	1.2
	Total trips	N/A	3.2
Philadelphia	Total trips	N/A	2.5
New York	Commute	11-23%	1.1-1.2
	Non-work	14-22%	0.6-1.7
	Total trips	N/A	1.7-2.9

N/A: not available.

Commuter Check programme in San Francisco found the programmes result in an increase in employee public transit use for both commuting and non-work trips among those receiving employer-provided public transit benefits (RSPA, 1995; MTC, 1995). As shown, employees receiving benefits took from 1.7 to 3.2 new public transit trips per week.

Although most of the employees receiving public transit benefits already commuted by public transit, the surveys suggest that most of the users who increased public transit use were previously non-users or infrequent users of public transit, and remain irregular users[1]. The largest increases in public transit use appear to be in suburban areas, where existing public transit share is lower than urban areas. For example, in the MTC study, the average increase was 3.0 new public transit trips for employees working in San Francisco and 3.7 new public transit trips per week for employees working outside San Francisco.

Other studies estimate higher values, around -0.3 for France and -0.5 for the UK for short-run elasticities with respect to fares. Litman (2004) also cites Gillen (1994) as demonstrating that car owners and users are (unsurprisingly) more sensitive to fare increases (*i.e.* other users are often "captive" to public transit), with a price elasticity of -0.41 compared to -0.28 for all users.

---

1. *It is not clear to what extent the level of the subsidy affects the number of new public transit trips. One would expect that a higher subsidy would yield greater public transit use. The San Francisco study, however, suggests that the level of the public transit subsidy has little bearing on the public transit ridership effect.*

Litman (2004) also finds that rail and bus elasticities often differ. This difference may be due to income differences, as higher-income residents tend to be more likely to use rail systems than buses. For example, Pratt (1999) estimated own-price elasticities of rail transport ridership to changes in transit fare in Chicago. Estimated elasticities were –0.10 and –0.46 for peak and off-peak riders, respectively, compared to –0.30 and –0.46 for bus riders. A study for the Australian Road Research Board (Luk and Hepburn, 1993) was cited by Litman (2004) as reporting average rail elasticities of –0.35, compared to –0.29 for bus.

Changes in public transit ridership do not translate directly into changes in private vehicle travel. Much depends on the particular circumstances of a transit system and the urban area in which it operates. For example, in many public transit free-fare zones, many of the patrons using the free public transit services likely would have walked or used public transit anyway in the absence of the free ride, thus resulting in limited private vehicle trip reduction. However, these programmes can still support vehicle trip reduction by increasing the likelihood that people will use them to get around for midday trips without a vehicle. This in turn could make car-pooling to work more attractive. Free services on commuter routes most likely will draw a much larger share of riders who otherwise would have driven to work and thus have much larger direct VMT reduction effects.

Litman (2004) cites Pratt (1999) as finding a range of 10% to 50% of increased trips by bus substituting for a car driver trip, while 20% to 60% of decreases in car driver trips will divert to public transit. Hagler Bailly (1999) estimated the breakdown of ridership sources for increased transit trips as 62% diverted from car trips, 4% from taxis and 34% from others such as cycling or walking. While Litman recommends using quite low short-term cross-price elasticities for car travel with respect to public transit fares (–0.03 to –0.10), this may understate mode switching during a situation such as a petroleum emergency as these estimates usually reflect a stand-alone *ceteris paribus* fare change.

Off-peak service enhancements

For most public transit systems, increasing service during an emergency would mostly be limited to off-peak periods. Typically, transit operators put their maximum available fleet in service during peak periods, constrained by rolling stock supply and a small reserve of vehicles to provide replacement for mechanical breakdowns or other operational contingencies. However, midday and other off-peak services can usually be increased significantly – though often at the expense of additional driver overtime and/or deferred regular maintenance usually conducted during this period.

Such service increases result not only in increased system capacity, but also increased service frequency and thus reduced traveller wait times. Many studies have shown that travellers place a high value on reducing wait times. For example, they typically put a higher value on reducing this "out-of-vehicle" wait time than on reducing "in-vehicle" time. Service increases could also provide better passenger comfort (less crowding), though this would depend on the overall response in terms of increased ridership.

Increasing bus service will lower a city's (and a country's) fuel demand by diverting trips from private cars. However, this strategy may have some offsetting effects in terms of increased demand for petroleum from public transit. Our calculations on the likely increase in bus fuel use compared to reductions in likely car fuel use indicate that this effect is probably negligible in most cases (see also DIW [1996], which did a similar calculation). This is mainly due to the large number of cars removed from the road for each bus added.

To determine the effect of improvements in public transit on ridership, two different types of studies should be consulted. In addition to empirical studies of the relationship between transit service level and ridership, travel modelling studies are useful sources of elasticity estimates. These studies, often conducted for particular urban areas, are able to roughly estimate increases in transit ridership and decreases in regional private vehicle travel from a wide range of public transit policies. Litman (2004) reviewed a variety of studies and concluded that, though there is considerable variation, the elasticity of public transit use with respect to public transit service frequency averages about 0.5. This elasticity relates the percentage change in transit trips to the change in "headway" time (the time between bus/train arrivals) or to out-of-vehicle wait time. Greater effects were found where transit service is infrequent.

Bus lane prioritisation enhancements

The third option for improving public transit is the creation or enhancement of dedicated lanes for service, such as bus lanes. While some communities have implemented grade-separated facilities (*e.g.* Ottawa, Pittsburgh and several Australian cities have roadways and highways dedicated to public bus service), more frequently these are on-street facilities, where only buses are allowed to use a particular lane or street. This is common in the United Kingdom and elsewhere in Europe. One strategy could be to extend the operational hours of bus lanes to 24 hours and weekends. Often, these facilities function as bus facilities only a few hours per day, usually one direction during peak rush hour. Changes in mode shares are highly dependent upon the travel time savings and reliability improvements that can be achieved by bus lanes.

The Urban Transport Industry Commission (1994) found bus demand elasticity with respect to bus in-vehicle time of about –0.7. A study by Hagler Bailly (1999) found a lower in-vehicle time elasticity for buses of –0.4, but also found this to be twice as big as the fare elasticity. This indicates that changes in the fare are not as important as changes in time when travellers choose between travel by bus and other means.

A report by the UK Commission for Integrated Transport assumes that, on average, trip times are reduced by 2.5% for every kilometre stretch of dedicated bus lane on the journey, compared to regular lanes. An average 10% time saving may be achievable if bus lanes cover half of the route (Commission for Integrated Transport, 2002a). These appear to be assumptions. Kain *et al.* (1992) report that central business district bus lanes in the United States increase bus speeds by up to about 25%, although this varies significantly depending on local circumstances.

One benefit of bus lanes is that creating them on-street is relatively cheap, requiring only road striping and signage. They can be set up quickly, though should be prepared in advance of fuel shortage emergencies.

Analysis of public transit policies

As described in the appendix, the Millennium Database contains detailed transport statistics for a large sampling of urban areas throughout the world. Those cities within the IEA countries with complete data were used in our analysis and are listed in the appendix (Table A-5). The same procedure used in the development of the database was used here for normalising public transit estimates to regional totals, starting from this sampling of urban areas. This relates data on total population for each region to total urban population for each region, and to the percentage of total urban population represented by the Millennium Database sample. While public transit ridership is likely to be disproportionately higher in the cities covered in the Millennium Database than in other urban areas, on the other hand this normalisation procedure does not account for rural, regional or short inter-city public transit services (*e.g.* many express commuter buses, etc.). A cross-check of these numbers against Eurostat figures for total bus and coach passenger-km showed comparable results at the regional level. Thus these data, and the approach for aggregating to regional totals, appear to provide reasonable estimates for baseline public transit ridership.

Public transit ridership data were available by mode from the Millennium Database. Ridership by mode, during peak and off-peak times, was used as the basis for estimating off-peak and weekend ridership at 45% of total ridership. Relevant data are shown in the appendix, Table A-6, and the results of the data normalisation are shown in Table A-7.

On the basis of the literature reviewed above, effectiveness factors and elasticities were selected for variants of each of the three policy approaches discussed above (fare reductions, service enhancements and lane prioritisation for transit). Two variants of each were chosen, for a total of six measures. These are shown in Table 2-8. For each measure, an elasticity was used to relate the change in fare or in time savings (from improved service) to a change in transit ridership. The elasticity calculations are shown in Table 2-8, with the impacts shown as percentage changes in daily transit trips. For the fare reduction measures, a cross-price elasticity impact on reduced private vehicle trips is also shown.

Table 2-9 shows how the impact estimates on transit trips were further developed, with calculations carried through to fuel savings. Just one of the six measures is shown: the 50% reduction in transit fares. A full set of estimates for all six transit measures is provided in tables A-10 and A-11 in the appendix.

Table 2-9 shows that, for the fare reduction measures, two different methods were used to estimate the reduction in private vehicle trips. First, a "diverted trips" measure was used. Based on the literature, 60% of the increased transit trips were assumed to have been "diverted" from private vehicles, and private vehicle trips were decreased accordingly. The second approach used the cross-price elasticity shown in Table 2-8.

Table 2-8

Elasticity assumptions and impacts for public transit measures

Measure	Impact	Estimation approach (type of elasticity used)	Japan/ RK	IEA Europe	US/ Canada	Aus/ NZ
Reduce public transit fares by 50%	Increase in transit trips	Apply own-price elasticity (–0.4 for Europe and Asia; –0.3 for North America and Australia/NZ)	20	20	15	15
	Decrease in private vehicle trips	Apply cross-price elasticity (–0.10) to private vehicle trips	–5	–5	–5	–5
Reduce public transit fares by 100%	Increase in transit trips	Apply own-price elasticity (–0.4 Europe and Asia; –0.3 North America and Oceania)	40	40	30	30
	Decrease in private vehicle trips	Apply cross-price elasticity (–0.1) to private vehicle trips	–10	–10	–10	–10
Increase weekend and off-peak service frequency by 40% (to peak levels)	Increase in transit trips	Apply out-of-vehicle time elasticity (0.5) to off-peak public transit trips	20	20	20	20
Increase off-peak service as above plus increase peak service frequency by 10%	Increase in off-peak/peak transit trips	Apply out-of-vehicle time elasticity (0.5) to off-peak/peak public transit trips	20/5	20/5	20/5	20/5
Convert all HOV and bus lanes to 24-hour bus priority usage	Increase in off-peak transit trips	Apply in-vehicle time elasticity (0.4) to a 10% average time saving on off-peak public transit trips	4	4	4	4
Convert as above plus designate two linear metres of new lanes per 1 000 urban residents	Increase in off-peak/peak transit trips	Apply in-vehicle time elasticity (0.4) to a 15% average time saving on off-peak public transit trips and 5% for peak trips	6/2	6/2	6/2	6/2

Note: positive numbers denote percentage increases in trips; negative numbers denote decreases.

Table 2-9

Estimated impacts of a 50% reduction in transit fares

	Japon/ RK	IEA Europe	US/ Canada	Australia/ NZ
Percentage increase in transit trips (own-price elasticity of -0.4 for Europe and Australia/NZ; -0.3 for other regions)	20	20	15	15
Additional transit trips per day (millions)	21.1	36.2	6.1	0.6
Reduction in private vehicle trips per day (millions)*				
• Method 1: apply 60% diversion factor to estimate private vehicle trips reduced	12.6	21.8	3.6	0.3
• Method 2: apply cross-price elasticity (-0.10) to private transport trips	7.5	22.0	37.5	3.0
• Final estimate (lesser of method 1 or 2)	7.5	21.8	3.6	0.3
Average private vehicle trip distance (kilometres)	12.2	12.4	13.2	9.9
Private vehicle reduction in daily travel (million kilometres)	91.8	269.7	47.5	3.0
Fuel saved per day (million litres)	10.2	27.3	6.8	0.5

*Note: for reduction in trips in private vehicles, results of two methods are shown in two rows; only the lower estimate is used in subsequent calculations.

As shown in Table 2-9, the lower of these two estimates was then selected as the more likely result and used for subsequent calculations. For all regions except Japan/RK, the "diverted trips" approach resulted in a much lower estimate of private vehicle trip reduction than the cross-elasticity approach, which in some cases yielded the implausible result of more car trips reduced than transit trips generated. Finally, Table 2-9 converts the daily trip reduction in private motor vehicles to reductions in daily vehicle-kilometres of travel (VKT) and fuel use.

Table 2-10 converts the daily fuel savings results into annual oil savings and the percentage this represents of total road transport fuel use and petroleum fuel use by region, if the policy were applied throughout the IEA.

An important caveat in these calculations is whether the assumed elasticities, estimated under normal conditions, are applicable for an "emergency" situation. On the one hand, fuel price increases and, especially, a fuel shortage may cause considerable shifting to public transit even without any measures.

Table 2-10

Estimated fuel savings from public transit measures: summary results

	Japan/ RK	IEA Europe	US/ Canada	Australia/ NZ	Total
Million litres saved/day					
50% fare reduction	10.2	27.3	6.6	0.4	44.5
100% fare reduction	20.4	54.7	13.5	1.0	89.5
Increased off-peak service	9.4	15.1	5.0	0.4	29.8
Increased peak and off-peak service	11.8	18.7	6.1	0.4	36.9
Bus and HOV enhancement	0.4	1.7	0.6	0.0	2.7
Bus and HOV expansion	0.8	3.4	1.1	0.1	5.4
Thousand barrels saved/day					
50% fare reduction	64.1	172.0	41.6	2.5	280.1
100% fare reduction	128.1	343.9	84.9	6.2	563.2
Increased off-peak service	58.9	94.9	31.2	2.5	187.5
Increased peak and off-peak service	74.3	117.4	38.1	2.5	232.3
Bus and HOV enhancement	2.6	10.7	3.5	0.2	16.9
Bus and HOV expansion	5.1	21.3	6.9	0.5	33.9
Percentage of road transport fuel saved					
50% fare reduction	3.1	3.0	0.4	0.5	1.4
100% fare reduction	6.1	6.1	0.7	1.2	2.8
Increased off-peak service	2.8	1.7	0.3	0.5	0.9
Increased peak and off-peak service	3.5	2.1	0.3	0.5	1.2
Bus and HOV enhancement	0.12	0.19	0.03	0.05	0.08
Bus and HOV expansion	0.24	0.38	0.06	0.09	0.17
Percentage of total fuel saved					
50% fare reduction	1.7	1.9	0.3	0.3	1.0
100% fare reduction	3.4	3.9	0.6	0.8	2.0
Increased off-peak service	1.6	1.1	0.2	0.3	0.7
Increased peak and off-peak service	2.0	1.3	0.3	0.3	0.8
Bus and HOV enhancement	0.07	0.12	0.02	0.03	0.06
Bus and HOV expansion	0.14	0.24	0.05	0.07	0.12

On the other hand, government measures to provide more and/or cheaper transit during an emergency may be seen as a valuable and genuine effort to alleviate mobility problems, and people may be more responsive than under normal circumstances.

Car-pooling policies

Encouraging car-pooling is another potential option for reducing private vehicle travel – by reducing single-occupant vehicle travel. Car-pooling refers to two or more individuals sharing a ride in a car, often on a regular basis[2]. Various policies for encouraging car-pooling have been devised. These include the construction of car-pool-only traffic lanes, preferential parking and methods for matching potential car-poolers (usually those commuting to the same place of employment).

Many cities in the United States have built car-pool-only lanes (also known as high-occupancy vehicle or HOV lanes) on major motorways, either by simply restriping existing lanes and adding signage indicating that the lanes are restricted to car-pools, or by major investment in new, dedicated roadway facilities (*e.g.* adding new lanes). Many areas in the United States find that the number of people carried in HOV lanes often exceeds those in regular ("mixed-flow") lanes, although most HOV lanes are still underutilised. Table 2-1 above shows estimates that car-pool lanes reduce total vehicle miles travelled by anywhere from 0.2% to 1.4%.

HOV lanes are less common in Europe, with only a few examples of dedicated facilities (Noland *et al.*, 2001). HOV lanes generally are found to be more effective when commute lengths are long (leading to greater travel time savings) or when commutes are to centralised zones, with high concentrations of employment and with easy access by public transit or on foot to other areas. HOVs usually consist of family members or friends. Several cities have found that many or even most HOV users are drawn from public transit when HOV lanes are constructed. However, under emergency conditions, many more HOV riders might be drawn from single-occupant cars.

Kuzmyak (2001) reported that the impacts of an HOV lane depend on numerous complex and interrelated factors. HOV lanes should certainly improve average traffic flow conditions in their own lanes, raising average speeds and reducing congestion. Depending on the degree of prior congestion

---

2. *The terms car-pooling, car-sharing, and ride-sharing are often confused. Car-sharing refers to the sharing of a car or cars by a group of people, taking turns. It also takes the form of car co-operatives and short-term rentals. Ride-sharing refers to the informal sharing of a ride (often between strangers) so that the driver can take advantage of car-pooling infrastructure (examples of this exist in San Francisco and Washington, DC). Car-pooling usually refers to more formalised, or at least more regular, ride-sharing arrangements. See Noland and Polak (2001) for more details.*

and success of the HOV lane in attracting ridership, flow on parallel lanes may be improved or worsened. In successful cases, HOV lanes provide higher levels of service (higher speed, reduced travel time) both for persons who previously drove alone and those who used public transit. In reviewing detailed regional studies, Kuzmyak found examples of HOV facilities both increasing and decreasing local air pollutant emissions. It is, however, unclear what the impact on fuel consumption would be, as this may not be correlated with impacts on emissions of other pollutants.

Noland and Polak (2001) summarise some of the issues involved with modelling HOV lane usage. Their report provides coefficient estimates used in regional models developed in the United States. McDonald and Noland (2001) developed a simple simulation model that uses others' estimated coefficients to evaluate changes in HOV usage. Their results suggest that travel time changes can generate shifts to HOV lanes; for example, an elasticity of –2.0 was found relating reduction in delay to incidence of HOV lane use. Table A-12 in the appendix displays the coefficients used by McDonald and Noland (2001) which were collected from a variety of sources.

Park and ride lots

Park and ride facilities are most commonly parking lots near freeway on-ramps or adjacent to regional transit or rail service. Park and ride lots allow for car-pool partners to meet one another and van-pool riders to meet at a central location, perhaps expanding the range in which ride-share arrangements can be efficiently formed. Many car-pools and van-pools use transit-oriented park and ride lots for this purpose. Parking lots at some rail stations prohibit parking by non-transit users.

The impact of park and ride facilities on car-pool and van-pool formation and use has not been widely evaluated. Clearly, such facilities, which often are located near freeways, make it easier for car-poolers to minimise the car-pool trip time. They also might provide a convenient place for van-pools to park overnight.

In one study of park and ride users in Dallas, 21% said they would not car-pool if it were not for the availability of the park and ride lot and 62% said the lot was one of many factors in deciding to car-pool. Another study of some 150 fringe car-pool park and ride lots showed that the prior mode of users was

60% single-occupancy vehicle and 34% car-pool, thus showing a substantial mode shift (Pratt, 1981). One issue related to park and ride facilities is whether they induce people to drive to a pick-up point rather than walk or be picked up at home, thus increasing vehicle travel and fuel use. No comprehensive evaluation has been conducted to date on car-pool and van-pool mode share increases due to enhanced marketing, promotion and education alone.

Various park and ride contingency plans could be developed for emergency fuel supply reductions, including locating and identifying existing parking facilities that could be converted to park and ride lots on short notice. These could include parking lots at existing shopping centres which may be underutilised much of the time. Rough estimates of the expected effectiveness of these types of lots could be made (based for example on the mode shift estimates discussed above).

Another element of park and ride lots is that they can encourage "casual car-pooling" or "informal ride-sharing". In Washington, DC, casual car-pooling occurs from the northern Virginia suburbs along the HOV (3 or more riders) lanes on I-395. Drivers pick up passengers at several locations (known as "slug lines") paralleling I-395, including a number of park and ride lots. The car-pools go to two destinations, the Pentagon in Arlington, Virginia and downtown Washington, DC This system is also used for evening trips, with slug lines forming in several locations. It is estimated that in the Washington DC area 2 500-5 000 commuters participate in a casual car-pool each day, mainly during peak travel periods (Noland and Polak, 2001). A similar system spontaneously appeared in the San Francisco Bay area, taking advantage of an HOV lane on the Bay Bridge, allowing morning queues of up to 30 minutes to be avoided.

Financial incentives for car-pooling

Financial incentives are public-sector supported programmes to reduce the cost of car-pooling, van-pooling or public transit, to increase the share of these modes. A wide range of different types of financial incentives can be offered, including: subsidies for van-pools; reduced public parking rates for registered car-pools or van-pools; special gasoline purchase cards or free carwashes for registered car-pools; and reduced-fare or fare-free public transit services. Special discounts can also be offered to car-pooling and transit commuters, such as receiving a discount at selected retailers, restaurants or services (*e.g.* such as oil change) for registered car-poolers or van-poolers or people showing a transit pass.

The theory behind financial incentives is that they will make it more appealing and less costly to car-pool or take transit and therefore encourage people to switch to these alternatives. Ample evidence suggests that commuters do respond to price signals. The effectiveness of these programmes depends on the type and level of the incentive.

Studies have concluded that financial incentives are a fundamental part of effective trip-reduction programmes. It has been estimated that incentives that give employees something extra, such as a subsidy, bonus or prize, can eliminate up to 20% of the daily vehicle trips arriving at work sites (Southern California Rideshare, 2003). According to a 1994 study of over one thousand Los Angeles area programmes for employee commute trip reduction, financial incentives were found to be the most effective of all the strategies evaluated (Cambridge Systematics, 1994).

Overall, the impacts of these ongoing financial incentive programmes depend on the type and level of incentive. For example, in many public transit free-fare zones, many of the patrons using the free transit services likely would have walked or used transit even without a free ride, thus resulting in limited direct vehicle trip reduction. Free services on commuter routes most likely will draw riders who otherwise would have driven to work and therefore could have larger direct vehicle travel reduction effects. Limited information is available on the effects of discounts and other small benefits for car-poolers. These programmes may have less effect on changing travel behaviour than they have on maintaining existing car-pools and van-pools.

A study for the Southern California Association of Governments, Regional Transportation Demand Management Task Force (LDA Consulting *et al.*, 2003) analysed two ongoing financial incentives: 1) a regional van-pool subsidy programme, providing $150 per month per van-pool; and 2) a regional car-pool incentive programme worth $25 per month per registered car-pool, which could be in the form of a prepaid gasoline card and/or discounts on public parking. The van-pool subsidy was estimated to reduce vehicle trips by 26 400 per day and vehicle-miles travelled by about 526 000 per day. The car-pool incentive was estimated to reduce vehicle trips by 137 000 per day and vehicle-miles travelled by 1.9 million per day.

The cost of financial incentives depends on the type of incentives offered. Direct government subsidies to registered car-poolers and van-poolers do require a substantial outlay of funds. Discount programmes that provide people who car-pool with savings at local stores may not require a government subsidy if local businesses offer discounts as a way to attract customers. Marketing, outreach and tracking, however, would be required in order to sign up businesses in the programme, raise awareness of the programme among travellers and make sure that participants are actually car-pooling or using public transit.

Although financial incentives are expensive, they are also usually very effective. People readily respond to price signals. Such programmes are likely to be more effective where there are supporting programmes, such as preferential parking and employer-based support for car-pooling.

Financial incentives require co-ordination with outside entities. In addition to the need to receive funding from local governments, these programmes will require co-ordination with van-pool providers, parking operators, etc., in order to be implemented. Any special discounts offered by retailers or related service providers also will need to be negotiated and publicised.

Each jurisdiction could operate a separate financial incentive programme, although it would be helpful if the basic programme elements and requirements were similar to avoid confusion. A preferable approach might be for the programme to be operated regionally or nationally.

These types of policies can be evaluated using the price elasticities discussed above. However, in a short-term emergency, these programmes would be difficult to set up and more direct policies might be preferable.

Car-pooling analyses

As previously discussed, policies to encourage car-pooling are quite broad. These have generally consisted of providing preferential car-pool lanes or parking spots for car-pools. Other policies, aimed at increasing the cost of single-occupant cars, often result in increased car-pooling (in addition to modal shifts to public transit). The analysis presented below focuses on the potential of car-pooling in an emergency situation. We assume that public appeals to car-pool, perhaps combined with preferential treatment of car-poolers (*i.e.* reserved lanes and parking) would lead to some increase.

Increased fuel prices during an emergency would also lead to some modal shift, but we do not explicitly analyse this.

Several estimates of the impact of car-pooling are calculated. First, to gauge an upper bound estimate, we make the simple assumption that every car trip now has one additional person who had previously driven alone. This is clearly an extreme assumption but serves as a high-end estimate of the maximum potential of car-pooling. We also estimate effects from adding one person to every car on a motorway, which could be a high-end scenario for large-scale, car-pool lane deployment. We also analyse adding one person to every car commute trip. Low-end estimates are calculated by analysing the effect of previous assumptions on VKT reduction from car-pool lanes (not under emergency conditions). Table 2-11 shows the intermediate estimates and results for adding one person to each car within all urban areas of each region, with the assumption that these extra passengers are drawn from existing single-occupant vehicles. Put another way, this assumes an increase in average car occupancy of one person.

We used the Millennium Database of cities (appendix, Table A-7) to estimate average vehicle occupancy for each region (Table 2-11). Note that our average vehicle occupancy numbers appear to be relatively high. This would, if anything, lead to less than expected reductions from this policy and thus this might represent a lower maximum potential than if current vehicle occupancy rates were lower.

Vehicle travel for each scenario was then recalculated using the assumed increase in vehicle occupancy, allowing us to estimate VKT saved per day and total barrels of fuel saved. Since these figures were calculated for only a sample of urban areas, we prorated this to cover all urban areas for each region and then further prorated to include the entire region (*i.e.* non-urban areas) using normalisation factors reflected in Table 2-12.

The basic formula for calculating this is:
Fuel saved = (total VKT) x (current average occupancy) / (new average occupancy) x (litres/km) / (percentage metro population in sample) / (percentage urban population in region)

Table 2-12 shows summary estimates for this case plus several other cases. As shown, our second estimate assumed that vehicle occupancy would only increase for motorway trips. This could be consistent with a policy of putting

Table 2-11

Impacts of adding one person to every urban-area car trip

	Japan/ RK	IEA Europe	US/ Canada	Australia/ NZ	Total
(Initial) average vehicle occupancy	1.50	1.37	1.40	1.53	
Daily urban VKT (millions) from Millennium sample of cities	529	830	1 964	203	3 526
Daily PKT (millions)	792	1 137	2 756	310	2 238
Daily VKT when adding one person to every car trip (millions)	318	479	1 148	123	2 068
VKT saved per day (millions)	211	350	817	80	1 458
Percentage VKT reduction	39.9%	42.2%	41.6%	39.4%	41.3%
Litres saved per day (millions)	24	36	114	11	185
Barrels saved per day (thousands)	148	224	715	67	1 154
Barrels saved per day, prorated for all urban areas (thousands)	289	977	2 560	134	3 960
Barrels saved per day, prorated for entire region (thousands)	363	1 233	3 320	158	5 073
Percentage of fuel used for transport saved, entire region	17.3%	21.9%	28.1%	30.0%	25.3%
Percentage of total fuel consumption saved, entire region	9.6%	13.9%	21.5%	21.3%	17.6%

Notes: PKT: passenger-kilometres travelled; VKT: vehicle-kilometres travelled.

car-pool lanes on motorways (or restricting them to car-pool use only). Data from OECD's International Road Traffic and Accident Database (IRTAD, 2004) were used to estimate the percentage of motorway mileage for each region. These data were not available for Canada or Australia/New Zealand, so percentage data from the United States were used as a proxy. European data also were not available for every country, so percentage figures are based on those countries with available data. Further details and intermediate calculations for this and the remaining car-pooling cases are shown in the appendix (Tables A-13 through A-16).

The third case presented assumes that one additional passenger is taken on every commute trip. This again assumes that all those extra passengers previously drove alone. Estimates are based on total commute VKT (calculated from employment estimates as shown in Table 2-15 in the discussion on

Table 2-12

Estimated car-pooling impacts under different circumstances

	Japan/ RK	IEA Europe	US/ Canada	Australia/ NZ	Total
Reductions in vehicle-kilometres of travel (as a percentage of total regional VKT)					
From adding one person to every car trip	17.8%	14.5%	15.2%	32.4%	15.8%
From adding one person to every commute trip	12.3%	8.0%	14.0%	12.7%	12.5%
From adding one person to every urban motorway trip	1.6%	3.3%	3.7%	7.8%	3.4%
From a 10% reduction in urban motorway VKT	0.4%	0.8%	0.9%	2.0%	0.8%
Thousand barrels saved per day, entire region					
From adding one person to every car trip	363	1 233	3 320	158	5 073
From adding one person to every commute trip	250	305	1 603	65	2 223
From adding one person to every motorway trip	33	277	800	38	1 149
From a 10% reduction in motorway VKT	8	66	192	10	276
Percentage of fuel used for transport saved, entire region					
From adding one person to every car trip	17.3%	21.9%	28.1%	30.0%	25.3%
From adding one person to every commute trip	11.9%	5.4%	13.6%	12.3%	11.1%
From adding one person to every motorway trip	1.6%	4.9%	6.8%	7.2%	5.7%
From a 10% reduction in motorway VKT	0.4%	1.2%	1.6%	1.8%	1.4%
Percentage of total petroleum fuel consumption saved, entire region					
From adding one person to every car trip	9.6%	13.9%	21.5%	21.3%	17.6%
From adding one person to every commute trip	6.7%	3.4%	10.4%	8.7%	7.7%
From adding one person to every motorway trip	0.9%	3.1%	5.2%	5.1%	4.0%
From a 10% reduction in motorway VKT	0.2%	0.7%	1.3%	1.3%	1.0%

VKT: vehicle-kilometres travelled.

telecommuting, below) and vehicle occupancy estimates for commute trips (which are lower than for all trips). No prorating of estimates is needed as these estimates are not based on the Millennium Database sample. Summary results are shown in Table 2-13.

Table 2-13

Consensus estimates of fuel savings from car-pooling

		Japan/ RK	IEA Europe	US/ Canada	Australia/ NZ	Total
Comprehensive policy of car-pool lanes, preferential parking and information systems	Barrels saved per day (thousands)	125	277	800	38	1 240
	Percentage road transport fuel saved	6.0%	4.9%	6.8%	7.2%	6.2%
	Percentage total fuel saved	3.3%	3.1%	5.2%	5.1%	4.3%
Policy to provide information and link ride sharers	Barrels saved per day (thousands)	13	41	112	6	171
	Percentage road transport fuel saved	0.6%	0.7%	1.0%	1.1%	0.9%
	Percentage total fuel saved	0.3%	0.5%	0.7%	0.8%	0.6%

Overall, these results show fairly large percentage reductions in fuel use, ranging from about 5% if all motorway trips have increased occupancy, up to about 22% if all trips do. If commute trips have increased occupancy, the estimated fuel savings is about 10%.

To put these results in some perspective, we can also use previous simulated estimates on how converting a motorway lane to a car-pool lane may reduce vehicle travel. As previously discussed, McDonald and Noland (2001) used travel time coefficients from a variety of models to estimate these effects over a 5 mile (8 km) corridor (Table A-12 in the appendix). VKT was reduced from 21.8 thousand to 19.6 thousand, or a 10% reduction in VKT on that corridor. With the estimation method described above, this works out to about a 40% reduction in VKT on motorways. This difference suggests that a lower-bound value from a car-pool lane policy might be based on a 10% reduction in motorway VKT. The impact on fuel savings is shown in Table 2-12 and is considerably lower than the maximum potential estimate. Since the 10%

reduction estimate is based purely on mode shifts due to travel time benefits during congested conditions, we would expect this to be a lower-bound estimate. Presumably, during emergency conditions, travellers would seek to car-pool to save both fuel and money, since we assume fuel prices would be higher; car-pooling would not occur just for travel time benefits.

Consensus estimate of reduction from car-pooling policies

The ability to increase car-pooling levels in an emergency is linked to the benefits that travellers see in choosing to car-pool and whether they are able to. For example, an extensive car-pool lane system will increase the travel time benefits of those choosing to car-pool as would priority parking measures. However, if large numbers of people car-pool, then traffic congestion would decrease, potentially encouraging more driving. These sorts of feedback effects may be less important during an emergency; when people will likely be more altruistic and will also respond because of fuel shortages and/or price increases.

In the analysis described above, we have presented a large range of potential effects. Table 2-12 reflects this range for different situations. It includes a fairly exceptional case (one more passenger in every vehicle for every trip) that would reflect an exceptional (probably altruistic-driven) response. Even assuming one extra passenger in all commuter or motorway trips may be optimistic, except under extreme circumstances.

The ability to increase car-pooling is clearly linked to both the circumstances (such as a normal situation versus an emergency) and the policies aimed at enabling increased usage. If a comprehensive set of policies were in place that, during an emergency, enable quickly invoking an extensive network of car-pool lanes, preferential parking facilities and good information systems for linking potential car-poolers, it seems reasonable that a high response rate could be achieved. If all major motorways were included in such a system, it seems reasonable that an average of one extra passenger per vehicle trip could be achieved, above what might happen without these policies. As shown in Table 2-13, this results in an average of a 4.3% reduction in fuel used for all IEA countries. However, the response rate for Japan and Korea is much lower than for other regions, due to the much lower rate of highway commuting. In this region, as a surrogate, half of all commute trips are assumed to add one rider (and a slightly different type of policy envisioned, less focused on motorways).

For a less ambitious policy effort, perhaps restricted to just providing information to encourage car-pooling and help potential car-poolers locate each other, our lower-bound impact estimate may be reasonable. This is the case shown in the lower part of Table 2-13, associated with a 10% reduction in motorway VKT.

Non-motorised travel and land use

Policies to increase the level of non-motorised modes of travel, such as walking and bicycling, have been pursued by various countries, especially in Europe, over the last 20 years. Recent evidence from Germany suggests that integrated policy approaches to increase the share of these modes can be quite successful. For example, Pucher (1997) reported that urban areas in West Germany have seen a 50% increase in the modal share of bicycle use between 1972 and 1995. He attributes this to a number of explicit policies undertaken to promote bicycle usage. These include building street infrastructure for bicycles (such as bike lanes) while making street networks more circuitous for cars. The latter is quite important, as a number of complementary policies have made it more difficult to use cars in urban areas, especially broad implementation of traffic-calming policies and parking-pricing policies. Many of these policies will also tend to increase pedestrian travel.

Many of these policies take many years to implement and to construct the networks needed to increase non-motorised travel. It is generally recognised that a broad policy package, not just construction of non-motorised facilities, is also needed to increase the cost of car use. No numerical estimates on the effectiveness of these policies have been found and most would need to be examined in combination with other measures. However, those areas with facilities in place would likely be able to more readily reduce fuel usage under short-term supply constraints.

Various land use characteristics such as the density of development, urban design and form, and the mix of uses have all been related to the propensity to travel by car. In particular, urban design can have a major influence on making areas more amenable for walking.

Those policies aimed at changing land development would obviously not have any short-run effects in emergency situations. However, understanding these effects can serve as a basis for understanding the flexibility of different countries and urban areas in their ability to respond to short-run fuel emergencies. We briefly summarise some of these issues and various estimates of the effectiveness of various changes in land use relationships. These estimates are largely drawn from a newly released Transit Cooperative Research Program report (TCRP, 2003).

Areas with higher population density typically will have lower rates of vehicle travel. This is partly due to increased proximity of destinations, but also to the higher relative cost of travelling by car and the increased provision of other modes. Separating out these influences to determine a "pure" density effect, TCRP (2003) reports results from Ewing and Cervero (2002) who derived elasticities of vehicle trips and kilometres travelled with respect to changes in population and employment density of –0.05. Their claim is that this can be added to other built environment or urban design factors, though it is not clear how public transit availability feeds into this relationship.

Measurement of land use mix (diversity of uses) tends to be more complicated than aggregate measures of density. Some of the variables described include measures of accessibility, entropy and dissimilarity of land uses. These require detailed spatial data to be fully characterised (TCRP, 2003). VMT elasticity estimates for each have been estimated at –0.3 for accessibility, –0.1 for entropy and –0.1 for dissimilarity.

Another critical land use issue concerns detailed site design characteristics. This includes details surrounding the existence of pedestrian linkages, such as sidewalks and street-crossing opportunities; street widths and block size; protection of pedestrians from street traffic, including aesthetics of the walking environment; set-back of buildings from the street and location of parking facilities. Again, the policies and measures needed to implement these sorts of changes take time to implement and have impacts. However, areas with beneficial site design characteristics may quickly show larger amounts of pedestrian activity, although estimating and generalising about total effects is difficult.

Street and town centre closures

One possible policy measure, normally aimed at increasing walking activity and that can be implemented relatively quickly, is the closure of some streets to motorised traffic. This is especially effective in the urban core areas of cities. The immediate impact is likely to be some reduction in vehicle travel, as documented by Cairns *et al.* (1998).

Closures of some urban centre roads have generally been implemented as part of pedestrianisation schemes, particularly in Europe. Implementation of many of these schemes was initially quite controversial, in that many feared that traffic congestion would increase significantly. What has been found, in many cases, is that some proportion of the traffic "disappears", that is, the demand is suppressed by the reduction in road capacity, with an increase in walking and cycling. This is essentially the opposite of what is commonly known as "induced demand", whereby the addition of road capacity can spur increases in motorised traffic (Noland and Lem, 2002).

Cairns *et al.* (1998) evaluated nearly 100 case studies of various reductions in road capacity, including street closures, to determine the impact on travel. They found that for most schemes there was a measurable reduction in traffic in the local area. They strongly caveat this result in that the actual effects are highly dependent upon the specific context and conditions within the local area. For example, this includes the availability of public transit, the type of parking controls in place, existing levels of traffic congestion and the overall walkability of surrounding areas. Another issue is one of measuring the effects. It is unknown how much of the "disappearing" traffic may actually be going to other roads or town centres where there are no restrictions, perhaps leading to a net increase in total traffic.

Despite these caveats, Cairns *et al.* conclude that, on balance, there is a net reduction in total traffic. However, estimating these effects would require detailed information on the local area. Their overall estimate is a 25% reduction in traffic relative to the original traffic on the affected streets. Thus, information on vehicle travel on those streets prior to the closures would be needed. Rough estimates can be based on current urban VKT.

Estimates of fuel savings from street closures

A very rough procedure was used to estimate the potential fuel savings from street closures in urban areas. This was to take the total vehicle-kilometres travelled for each urbanised area and the total kilometres of road in each urbanised area. Making the assumption that VKT per road length would be uniform throughout an urban area, a simple ratio of these values was used to estimate total VKT per km of road. This is clearly a very general assumption, as there will be tremendous variation in the utilisation of road space throughout an urban area. However, most congested areas will tend to have a greater than average level of road utilisation, so this assumption should result in a conservative estimate of the fuel savings effect. This ratio is then simply applied to determine a corresponding reduction in VKT as follows:

Fuel savings = % reduction in road length x (VKT/road length) x
(litres/km) x (estimate of disappearing traffic)

To estimate results, a sampling of cities was taken from the Millennium Database. Normalisation to regional totals was applied as previously discussed. For this measure, 2% of urban road space is assumed to be closed, with a resulting 25% reduction in traffic from that road space (*i.e.* 75% of the

Table 2-14

Estimated impacts of closure of 2% of urban road space

Regional averages	Japan/ RK	IEA Europe	US/ Canada	Australia/ NZ	Total
Percentage VKT reduction for sampled cities	0.5%	0.5%	0.5%	0.5%	
Daily VKT reduction for sampled cities (millions)	2.6	4.1	9.8	1.0	17.6
Percentage of total metro-area population	51%	23%	28%	50%	
VKT reduction prorated to all urban areas (millions)	5.2	18.0	35.2	2.0	60.4
Litres saved per day (millions)	0.6	1.8	4.9	0.3	7.5
Barrels saved per day (thousands)	3.6	11.4	30.8	1.7	47.5
Percentage of fuel used for transport saved, entire region	0.2%	0.2%	0.3%	0.3%	
Percentage of total fuel consumption saved, entire region	0.1%	0.1%	0.2%	0.2%	

traffic shifts to nearby roads). Our results suggest that this type of policy will only play a minor role in reducing regional road transport fuel consumption, around 0.2%, as shown in Table 2-14. If we assume effects are proportional to scale, 10% of urban area streets would need to be closed to yield a 1% reduction in regional road transport fuel use.

Thus, while these types of policies might be effective on a local scale, they are unlikely to show much impact unless implemented on a very large scale. Their effectiveness may also be enhanced when done in combination with other policies, such as increasing public transit service.

Work-trip reduction policies

Several policy options are focused on reducing the number of commute trips needed for individuals to engage in work activities. This includes policies to encourage more home-based work (also known as telework or telecommuting) and flexible work schedules. These types of policies generally can be implemented by employers. Various government policies can be used to encourage employers to adopt these types of policies.

One of the arguments sometimes used against increased home-based work or flexible work schedules is concern about management of employees. Also, employers may often feel the need to have all employees at the office at certain times, so that communication and worker interactions can be facilitated. While some of these concerns may be important for long-term changes to work habits, under emergency conditions, we expect these concerns could be temporarily set aside for many more employers.

Telecommuting or working at home

Telecommuting can be strictly defined as working at home but maintaining office contact via telecommunications. This contact can be either through the phone or computer. Essentially, this term is used for home-based office work.

Though many studies of telecommuting have been conducted, the net impacts of telecommuting on fuel consumption are still uncertain and difficult to estimate. This is due to uncertainty and variation in how telecommuters behave. For example, while they may avoid peak travel during congested conditions, it is unknown how much additional travel they may make from home during the day (such as shopping trips) which they would not have otherwise made. Long-term telecommuters may subsequently tend to relocate to live further from their workplace than non-telecommuters. Induced travel effects may also reduce the benefits of removing telecommuters from peak traffic flows (US DOE, 1994). However, from a short-term perspective under fuel shortage conditions, telecommuting can offer some fraction of the workforce the opportunity to continue to engage in economic activity without travelling to work. Thorpe *et al.* (2002) reported a significant rise in telecommuting during the British fuel emergency, although the absolute numbers were small.

To estimate these effects we need to broadly understand which segments of the workforce can potentially telecommute and what their average commute trip currently is. The United States Department of Energy (US DOE, 1994) reports projections of between 7.5 and 15 million telecommuters by 2002, representing some 5.2-10.4% of the workforce. About half of these would commute to local centres which provide facilities for telecommuters, so their work-trip vehicle travel would not be completely eliminated. Average round-trip commute length was estimated at 21.4 miles (34.5 kilometres) with commutes to telecommuting centres being a round trip of 9 miles, resulting in about 12.4 miles of reduction in their commute trip. Estimated total VKT reductions and the number of telecommuters are shown in Table 2-15.

While estimates of the number of telecommuters vary, US Department of Labor statistics released in 1999 suggest that approximately 12% of the workforce telecommuted occasionally. Some regions of the country, such as Washington, D.C., have higher telecommuting participation. The Washington, D.C. region "State of the Commute Survey" conducted in 2001 showed that 15% of commuters telecommuted either regularly or occasionally and another 18% of commuters said their job responsibilities would allow them to telecommute and they would do so if their employer permitted it. (WMCG, 2002).

Evidence suggests, however, that some extra travel on telecommute days will occur. For example, the Telework America 2000 (ITAC, 2000) survey found that about 25% of the reduction in commute vehicle travel was offset by increased travel for errands on telework days[3].

---

3. Telework America 2000 *found an average of 4.5 to 5.0 extra miles travelled on teleworking days for errands. This compares to an average commute distance for teleworkers of 19.7 miles.*

Table 2-15

Projections of telecommuters and telecommuting miles

Telecommuters	Actual 1988	Projected 2005	Projected 2010
Information workers as % of all workers	54.8	60.0	61.1
Telecommuters as % of information workers	1.3	27.8	44.9
Telecommuters as % of all workers	0.7	16.7	27.4
Total number of telecommuters (millions)	0.5	17.7	29.1
Vehicle-kilometres avoided by telecommuting (billion)	1.8	66.3	108.7

Source: US DOE (1994).

Short-term fuel consumption savings could be estimated using information on telecommutable jobs in the economy. Average work travel distances would also need to be known and average fuel consumption per mile.

Specific policies can be pursued to promote telecommuting, especially under emergency conditions. Persuading employers that telecommuting would not be harmful may be necessary. This can be done by educating employers about the potential costs and benefits. One possible policy mechanism is to sign up large employers to a telecommuting programme that would be implemented under emergency conditions. Employers would agree to have certain employees telecommute at least part of the time during any fuel shortage emergency.

Some infrastructure might be needed to allow employees to work at home. At a minimum, some employees may simply need a home computer, although many of those who would have jobs conducive to telecommuting may already have home computers. Internet access may also be needed and employers may want to pay for broadband access, if needed. Again, most employees with telecommutable jobs may already have modem-based Internet access, which might be suitable for the majority of work needs.

Telecommuting analysis

Estimates were made of the maximum possible savings in fuel consumption due to telecommuting. These estimates were made for the four IEA regions using country-specific data, with various averages assumed for countries and regions lacking data.

The basic method was to first estimate the fraction of total jobs that could potentially be "telecommutable" or where work could be done at home. Not all jobs that can be done at home require information or communications technology and an attempt was made to classify jobs accordingly.

The key variables to consider are:

E = Total employment

TE = Total number of employees who could feasibly telecommute or work at home for a short period of time (who are not doing so already)

L = Average commute trip length, one-way (km)

C = Modal share of commute trips currently done by car (%)

R = Average car occupancy rate

F = Average fuel intensity of vehicle fleet (litres/100 km)

This leads to a simple equation for estimating the maximum potential level of telecommuting:

$$\textit{Maximum telecommuting fuel savings} = \frac{TE \bullet L \bullet 2 \bullet C \bullet . F}{R \bullet 100} \quad \textit{(litres)}$$

Data on total employment by job title were available for the United States. While these data might also be available for other countries, we were not able to locate them at this time. Therefore, maximum estimates of telecommutable jobs are based on United States data. Other estimates of telecommuting are often based on the fraction of service sector jobs. While many of these are not necessarily telecommutable, most of the IEA countries have between 60% and 70% of jobs in this sector, which is comparable to the estimates derived below.

A dataset was obtained from the US Bureau of Labor Statistics listing over 25 000 job titles by industry category and total employment for each. This list was examined to eliminate job categories that could not be engaged in at home. The resulting estimate was that 58% of all employees could potentially telecommute or work at home at least some of the time. This served as the basis for our maximum percentage telecommuting estimate for all the countries and regions analysed. Interestingly, this figure is comparable to the estimates of nearly 60% of the economy being "information" workers, as shown in Table 2-15.

Average data for length of commute trips and modal split for each region were based upon data from the Millennium cities (Table A-7 in the appendix).

While these estimates exclude rural populations, we expect that the vast majority of employees within the job categories selected would work in urban areas. Mode splits are also based upon urban area averages and most likely higher car usage rates would occur in rural areas. Average vehicle occupancy rates are relatively low for commute trips. Estimates of current telecommuting among those employees able to telecommute were based on the assumption that they telecommute twice a week and that 28% of potential telecommuters already do so. Other assumptions are listed in Table 2-16 along with estimated maximum reduction in vehicle-kilometres travelled (VKT) for each region.

The resulting estimates (in Table 2-17) show a maximum potential fuel savings of 1.8 million barrels of gasoline per day for those countries analysed. This total is based on the assumption that total telecommuting take-up is 100% in the event of a fuel supply emergency. More realistically, some fraction of this total would telecommute. United States estimates are that between 28% and 45% of information technology workers would engage in telecommuting.

Table 2-16

Input values and estimation of maximum VKT savings from telecommuting

	Japan/ RK	IEA Europe	US/ Canada	Australia/ NZ
Average commute length (km)	14	9	17	13
Percent private car trips	42%	49%	86%	79%
Total employed (millions)	85.0	133.0	144.6	8.4
Estimated share of employed who could telecommute	58%	58%	58%	58%
Potential telecommuting employees (millions)	49.5	77.4	84.3	4.9
Average commute trip vehicle occupancy	1.25	1.15	1.10	1.10
Total commute (million VKT/day)	804	1 025	3 846	162
Maximum potential savings (million VKT/day)	469	598	2 242	95
Estimated current savings from telecommuting (million VKT/day)	52	66	249	10
Estimated additional maximum savings (thousand VKT/day)	417	531	1 992	84

Table 2-17

Potential fuel savings from telecommuting

	Japan/ RK	IEA Europe	US/ Canada	Australia/ NZ	Total
Thousand barrels saved per day					
Telecommute every day					
Maximum potential fuel savings (all regions), 100% take-up	219	255	1 308	53	1 835
Low estimate, 25% up-take	55	64	327	13	459
High estimate, 50% up-take	109	127	654	26	916
Telecommute only twice per week					
Maximum potential fuel savings (all regions), 100% take-up	88	102	523	21	734
Low estimate, 25% up-take	22	25	131	5	183
High estimate, 50% up-take	44	51	262	11	368
Percentage of road transport fuel saved					
Telecommute every day					
Maximum potential fuel savings (all regions), 100% take-up	10.4%	4.5%	11.1%	10.0%	9.1%
Low estimate, 25% up-take	2.6%	1.1%	2.8%	2.5%	2.3%
High estimate, 50% up-take	5.2%	2.3%	5.5%	5.0%	4.6%
Telecommute only twice per week					
Maximum potential fuel savings (all regions), 100% take-up	4.2%	1.8%	4.4%	4.0%	3.7%
Low estimate, 25% up-take	1.0%	0.5%	1.1%	1.0%	0.9%
High estimate, 50% up-take	2.1%	0.9%	2.2%	2.0%	1.8%
Percentage of total fuel saved					
Telecommute every day					
Maximum potential fuel savings (all regions), 100% take-up	5.8%	2.9%	8.5%	7.1%	6.4%
Low estimate, 25% up-take	1.5%	0.7%	2.1%	1.8%	1.6%
High estimate, 50% up-take	2.9%	1.4%	4.2%	3.6%	3.2%
Telecommute only twice per week					
Maximum potential fuel savings (all regions), 100% take-up	2.3%	1.2%	3.4%	2.9%	2.6%
Low estimate, 25% up-take	0.6%	0.3%	0.9%	0.7%	0.6%
High estimate, 50% up-take	1.2%	0.6%	1.7%	1.4%	1.3%

Note: includes 25% increase in non-work driving.

If we therefore assume that, on average, between one-fourth and one-half of all potential telecommuters do so under emergency conditions, then the total savings would be between about 630 to 1 260 thousand barrels per day (not shown in the table). Another potential factor is that some telecommuters may increase non-work trips. Under normal circumstances, regular telecommuters in the United States have been estimated to offset their telecommuting miles by about 25% with other new trips. While this effect may be smaller under emergency conditions, we use this rough figure to get an estimated net reduction of about 450 to 900 thousand barrels per day for five-days per week telecommuting. For a case where people telecommute on average only 2 days per week, an estimated reduction of 190 to 380 thousand barrels per day results. Potential fuel savings range from about 1% to 6% of total fuel use across the IEA and up to 9% in some regions.

Consensus estimate of reduction from telecommuting policies

The analysis above shows a range in the effectiveness of telecommuting for reducing fuel consumption. The actual impact would likely be related to the type of policies put in place prior to any emergency, affecting how easily employees can work from home. Policies could include making sure that all employees (who need them) have both a computer and a broadband connection to the Internet. Probably more critical is to obtain the commitment of employers so that they allow their employees to work from home at least several days a week, during an emergency. Since many workers may not be able to purchase fuel under emergency conditions, some initiative from employers seems likely.

In our consensus estimate, we assume that all employees who can telecommute do so twice a week. This would represent an average value, as some might telecommute every day, while others might not telecommute at all. (Those who already telecommute some of the time, excluded in our estimates, might also increase their telecommuting frequency.) Though under emergency conditions it is less likely that large increases in non-work travel would offset these reductions, we assume an average 25% increase in non-work driving. The consensus estimate (based on the above analysis) is presented in Table 2-18 and assumes that employers are supportive of telecommuting and that they have provided resources to their employees (computer and Internet connection) to make it possible. Clearly, many employees may already have a computer and so the costs of implementation

Table 2-18

Consensus estimate of effect of telecommuting measure

	Japan/ RK	IEA Europe	US/ Canada	Australia/ NZ	Total
Thousand barrels saved per day	88	102	523	21	734
Road transport fuel saved (%)	4.2	1.8	4.4	4.0	3.7
Total fuel saved (%)	2.3	1.2	3.4	2.9	2.6

would range from zero to whatever investment level employers feel they need to make to facilitate telecommuting. This is discussed further in the cost/benefit analysis in Chapter 3.

Flexible/compressed work schedules

Flexible and compressed schedules, sometimes described collectively as "alternative work schedules", allow employees to work a full-time work schedule in arrangements other than the conventional five days per week, 7-8 hours per workday. Compressed schedules allow employees to work fewer days per week but longer days. In the United States, typical compressed work schedules are a 4/40 work week (working four 10-hour days per week with one weekday off every week), a 9/80 work week (working eight 9-hour days, one 8-hour day every two weeks, with a day off every other week), and a 3/36 work week (working three 12-hour days per week with two weekdays off each week). In countries like France, with fewer working hours per week, a 4/35 system where employees work three 9-hour days and one 8-hour day may be possible.

Flexible schedules do not change the length of the average workday, but allow employees to choose their start and end times, usually around a set of "core hours," during which time all employees are working. For example, if the core hours are from 10 a.m. to 3 p.m. and the workday is 8 hours long, one employee could choose to work 7 a.m. to 3 p.m., while another works 10 a.m. to 6 p.m. Programme efforts to encourage flexible or compressed schedules include providing information, technical assistance and financial incentives to employers to help them adopt and manage a programme.

According to the Southern California Association of Governments State of the Commute Report (SCAG, 1999), 32% of surveyed commuters said their employers offer some form of compressed schedule and about 16% of these employees participate, representing about 5.4% of the total regional population. Of these schedules, the 4/40 work week is most popular; 18% said their employer offers a 4/40 work week, and of these 12% participate. Nine percent reported that their employer offers a 9/80 work week, and of these 29% participate. Five percent of area commuters said their employer offers a 3/36 work week, and of these 12% participate.

The theory behind flexible schedules is that some employees who want earlier or later work schedules will shift their commuting time to off-peak hours, thereby freeing peak-period road capacity. In general, they have not been shown to reduce trips in substantial numbers. Thus, there is no vehicle travel reduction benefit and the fuel reduction benefit is limited to possible savings from reduced congestion. However, they can spread demand for public transit over a longer peak, thus allowing public transit to operate more effectively during emergency conditions.

Compressed work schedules result in the elimination of some work trips altogether and shift remaining trips to earlier or later travel times. For this reason, compressed schedules are more useful than are flexible schedules for meeting fuel consumption reduction goals. Some research, however, indicates that the actual trip and VKT reductions from compressed work weeks might be more modest than these calculations would suggest, because some participants make additional trips during their non-workdays (Giuliano, 1995).

During emergency conditions, this type of policy could be relatively easy to implement and employer co-operation would likely be greater. Some commuters, however, may have inflexible schedules dictated by other activity commitments (*e.g.* child-tending) making it difficult for them to alter their schedules quickly. Data on the type of jobs and demographic sectors that are most amenable to flexible schedules would allow rough estimates of the fuel savings to be calculated from the amount of uptake by individuals (employers could perhaps make it compulsory, although probably most countries would need legislative changes to make this legal under emergency conditions).

Compressed work week analysis

The potential of compressed work weeks is analysed in terms of their fuel saving potential. Work weeks of 4/40 and 9/80 are analysed, which correspond to a 20% and 10% drop in VKT of the employee, respectively. We assume that no new

Table 2-19

Results for compressed work week measure: 4 days/40 hours

	Japan/ RK	IEA Europe	US/ Canada	Australia/ NZ	Total
Fuel savings (thousand barrels per day)					
All employees	105.4	122.6	629.5	25.5	883.0
32% uptake	33.7	39.2	201.4	8.2	282.5
Telecommutable job uptake	61.4	71.5	366.9	14.8	514.7
Road transport fuel reduced (%)					
All employees	5.0	2.2	5.3	4.8	4.4
32% uptake	1.6	0.7	1.7	1.5	1.4
Telecommutable job uptake	2.9	1.3	3.1	2.8	2.6
Total petroleum fuel reduced (%)					
All employees	2.8	1.4	4.1	3.4	3.1
32% uptake	0.9	0.4	1.3	1.1	1.0
Telecommutable job uptake	1.6	0.8	2.4	2.0	1.8

Table 2-20

Results for compressed work week measure: 9 days/80 hours

	Japan/ RK	IEA Europe	US/ Canada	Australia/ NZ	Total
Fuel savings (thousand barrels per day)					
All employees	65.0	75.3	388.7	15.8	544.9
32% uptake	20.8	24.1	124.4	5.1	174.4
Telecommutable job uptake	37.9	43.9	226.6	9.2	317.6
Road transport fuel reduced (%)					
All employees	3.1	1.3	3.3	3.0	2.7
32% uptake	1.0	0.4	1.1	1.0	0.9
Telecommutable job uptake	1.8	0.8	1.9	1.8	1.6
Total petroleum fuel reduced (%)					
All employees	1.7	0.9	2.5	2.1	1.9
32% uptake	0.6	0.3	0.8	0.7	0.6
Telecommutable job uptake	1.0	0.5	1.5	1.2	1.1

trips are generated during the days one does not work, which seems a reasonable assumption during emergency conditions and since the total number of commute trips drops by a small amount. Our starting point for VKT is based on calculations from input values shown in Table 2-16 for the telecommuting analysis. We also assume that those currently telecommuting continue to do so and this VKT is subtracted from the total commute VKT.

Three possible scenarios are considered. One is that all employees participate in a compressed work week schedule. This provides an upper-bound estimate on the potential fuel savings from this sort of policy. The second assumes a 32% participation rate amongst employees, based on results reported above for Los Angeles (SCAG survey). The last calculation assumes that those jobs that are "telecommutable" as defined above are those that can easily engage in a compressed work week (this is about 58% of total employment).

Results are shown in Tables 2-19 and 2-20. The maximum potential petroleum fuel savings across the IEA is about 3% for the 4/40 compressed work week and about 2% for the 9/80 compressed work week. Other estimates are proportional to the participation rate, with about a 1% total fuel savings if only 32% of employees participate in a 4/40 programme.

Consensus estimate of reduction from compressed work week policies

The analysis above has shown several possible scenarios for compressed work week policies. It is unlikely that all employees would have schedules that make this possible even if all employers allowed it. The 32% uptake measured by some studies serves as a lower-bound estimate of all employers offering a compressed work week option. For an aggressive programme, a more reasonable assumption is that the type of jobs that are telecommutable (about 60%) would be a best estimate. We use this in our consensus estimate summarised below, assuming that a 4/40 policy is adopted. The key policy needed here is a requirement that employers allow their employees to work compressed schedules during a fuel emergency. Results are shown in Table 2-21.

Regulatory approaches to traffic reduction

Several regulatory approaches that expressly forbid traffic from certain areas or times of day are another potential policy mechanism for saving

Table 2-21

Consensus estimates for 4/40 compressed work week measure

	Japan/ RK	IEA Europe	US/ Canada	Australia/ NZ	Total
Fuel savings (thousand barrels per day)	61	71	367	15	515
Road transport fuel saved (%)	2.9	1.3	3.1	2.8	2.6
Total fuel saved (%)	1.6	0.8	2.4	2.0	1.8

fuel. These types of policies range from closing specific streets or town centres to traffic (as already discussed above), to mandated "car-free" days that forbid anyone (usually with some exceptions) from driving. The latter are often implemented as "odd/even" driving bans, such that those with licence plates ending with either an odd or even digit are banned on alternating days, or some variant thereof. Weekend driving restrictions are also often discussed. Experience with these approaches and rough estimates of their effectiveness are discussed.

Driving bans

Odd/even day driving bans were first discussed as a potential policy measure during the oil crises of the 1970s. While not used in many countries to restrict driving, they were instead used as a means to regulate queues at gasoline stations. This was done by selling gasoline to only those vehicles with their licence plate ending in an even or odd number on corresponding days of the month. While this did help reduce queueing and excess fuel consumption from idling while in the queue, it is unlikely to have had more than a very small effect on total driving and fuel consumption, since only queueing and not driving were banned.

Odd/even day driving bans have more typically been implemented as a means of reducing central city air pollution. Athens and Mexico City are the two cases most frequently discussed. While this policy is seen as generally effective over short time periods, eventually people tend to find ways of evading the intent of the policy. This is done primarily by acquiring or otherwise gaining access to a second vehicle with a complementary licence plate number. The policy also encourages people not to dispose of older vehicles in order to keep two vehicles available to the household, which is particularly counter-productive from an air pollution perspective (as older vehicles would tend to pollute much more than newer ones).

A good example of a short-term (one-day) implementation of driving bans was during the Paris air pollution emergency of 1997. Estimates are that total traffic and fuel use were reduced by about 30%. Air quality improved dramatically (although it is unclear how much this was due to meteorological conditions or the driving ban) and the ban was discontinued on the following day.

These types of policies are likely to be effective (and more politically acceptable) in emergency conditions when people are aware of the need to take strong actions. Their effectiveness will also depend upon the availability of other travel options (such as public transit or car-pooling opportunities) since if people do not have other options, they are more likely to drive despite the ban. Thus any evaluation of the potential fuel savings needs to consider these elements. In addition, the prevalence of households with more than one vehicle will also have an impact on both the feasibility and effectiveness of these policies. For example, with increasingly large segments of the population having two or more vehicles in the household, odd/even day driving bans may become much less effective than they were 10 or 20 years ago. This will also depend on where people live and work, and the feasibility of sharing rides to work. Some shared rides may also be longer if trips are made to drop someone off at a destination.

A rough assessment of these effects can be estimated with information on public transit availability, household car ownership rates and employment rates of households. Any estimate of these effects is likely to be based on rough assumptions, as there is little knowledge of the potential behavioural effects that may take place over time. In the extreme case of Mexico City and Athens, long-term implementation of the policy has been completely ineffective. On the other hand, the short-term (one-day) implementation in Paris appears to have been highly effective. Whether this would persist over a period of several weeks to a few months is unknown. Careful enforcement may be especially important over longer periods.

Estimates for various driving bans in Germany are shown above in Table 2-2. These are based on assumptions about how driving behaviour would be affected and therefore may not represent actual behavioural reactions. Some additional assumptions used in the German study are shown in Table 2-22.

Table 2-22

Assumptions used in DIW study for Germany

	Leisure VKT	Education VKT	Work VKT	Business VKT
General ban on Sunday driving	90% reduction	95% reduction, 5% shift	10% reduction, 40% shift	5% reduction
Alternate ban on Sunday driving (every other week)	65% reduction	95% reduction, 5% shift	10% reduction, 40% shift	5% reduction
	Saturday VKT		**Sunday VKT**	
General ban on weekend driving	70% reduction		85% reduction	
Alternate ban on weekend driving (every other week)	50% reduction		65% reduction	

Source: DIW (1996).

These show percentage reductions in vehicle-kilometres travelled (VKT) for various trip purposes and overall totals for weekend days. This study also assumed that a small fraction of VKT will be shifted to other days for some types of trips.

Some of the VKT reductions assumed in the DIW study may be low, if many people are creative about alternative approaches to their activities (such as car-pooling, shifting days of the week, etc.). It is likely that this type of policy would result in better trip planning, such as increased trip chaining to engage in more activities on a given day. However, current knowledge is insufficient to reasonably estimate these effects.

A more effective policy than weekend driving bans would be an "odd/even" or one-day-a-week ban tied to licence plate numbers. One method of estimating these effects would be to consider the distribution of the number of vehicles owned by each household. This would enable one to estimate the likelihood that no car is available to the household on a given day. The more vehicles available to the household, the less likely that this sort of policy will affect non-work trips and VKT for those trips. We would expect some small reduction in VKT, as households would then tend to pool their non-work activities. Perversely, this sort of policy would require the household to perhaps use a less fuel-efficient vehicle for these types of trips, if the more efficient vehicle is banned on that day.

The impact on work trips will depend on many factors. First, if alternative opportunities are available (such as public transit, car-pooling or even telecommuting), then this could be quite effective. If these modes are not available, then households with more than one car may actually increase their work-related VKT as more circuitous trips may be made to drop off and pick up one of the workers.

Analysis of driving bans

Given these complexities, we estimate some rough scenarios on the potential fuel saving effects of odd/even day driving bans. Our simplest estimate examines the impact of an odd/even day driving ban with the assumption that 50% less VKT will be generated. This leads to a 50% reduction in fuel consumption, which is very unlikely to occur, but could potentially indicate the maximum potential for this type of policy. An alternative driving ban of one day in ten (based, for example, on when the last digit of the vehicle licence plate matches the last digit of the day of the month) would give a maximum reduction of 10% (neglecting the details of months with only 30 days or less).

A more realistic estimate considers the structure of household vehicle ownership. The more vehicles that a household owns, the less likely it is that this sort of policy will have a major impact on a given household. For example, a two-vehicle household has a 50% probability that one vehicle cannot be used on a given day. This leads to a 25% probability that the household will not have a vehicle available every other day. A three-vehicle household would only have a 12.5% chance of all vehicles having odd or even licence plates and thus not having a vehicle available.

This is calculated simply as,

$$P = B^n$$

Where P is the probability of a vehicle being available to the household, B is the percentage of vehicles banned on a given day (e.g. $B = 0.5$ for a 50% ban) and n is the number of vehicles owned by a given household.

In our estimates we assume that all trips previously taken are made if vehicles are available. We assume no increase in driving from delivering people who do not have a car. But we also assume no shift in mode for those individuals in a household who may not have a car on a given day. These values are based on the distribution of the number of vehicles by household. Data for this were found for the United Kingdom for 2001 and the San Francisco Bay area for 1990. We use the distribution for the suburban areas of San Francisco to be representative of North America and the UK distribution for IEA Europe, Japan/RK and Australia/NZ. Interestingly, the United Kingdom distribution is very similar to that for the City of San Francisco, which is relatively less dependent on cars than most North American cities. The distributions are shown in Table 2-23, including an adjustment that excludes zero-vehicle households. As can be seen, the offset to the maximum potential is largest for North America, where the number of vehicles per household is generally larger than in other regions.

Table 2-23

Distribution of vehicle ownership by household

	City of San Francisco (1990)	Bay area excluding City of San Francisco (1990)	Distribution without zero-vehicle households	UK data (2001)	Distribution without zero-vehicle households
Zero vehicle	30.7%	7.4%		27.0%	
One vehicle	41.6%	32.5%	34.5%	44.0%	60.3%
Two vehicles	21.1%	3.9%	41.4%	23.0%	31.5%
Three or more vehicles	6.6%	22.6%	24.1%	6.0%	8.2%

A further adjustment is made that assumes that all VKT associated with work trips are still made. This could represent people being driven to work or dropped off *en route* to another destination. The simple estimate assumes that work-trip VKT is not reduced. Our assumption takes into consideration that while some trips may be shifted to other modes, other trips may actually be increased, such as a circuitous trip to drop someone off at work.

The calculated offset to our maximum estimate is shown in Table 2-24 for odd/even day bans and Table 2-25 for one-day-in-ten bans. As can be seen, the total offset is greatest for North America, which has the largest average vehicle ownership per household.

Total fuel savings, incorporating each of the assumptions, are shown in Tables 2-26 and 2-27 for each of the driving ban policies. These results clearly show that the maximum potential of this type of policy is unlikely to be

Table 2-24

Estimate of VKT reduction and offsets with odd/even day ban
(billion VKT and percentages)

	Japan/ RK	IEA Europe	US/ Canada	Australia/ NZ
50% reduction applied to all VKT	1.5	4.2	6.6	0.3
Adjust for vehicle ownership	1.1	3.3	4.0	0.2
Assume all commute VKT still made	0.7	2.7	2.1	0.2
Offset to maximum savings	21.9%	21.9%	38.8%	21.9%
Offset with all commute VKT still made	49.5%	34.2%	68.1%	48.6%

Table 2-25

Estimate of VKT reduction and offsets with one-in-ten-day driving ban
(billion VKT and percentages)

	Japan/ RK	IEA Europe	US/ Canada	Australia/ NZ
10% reduction applied to all VKT	0.29	0.83	1.31	0.06
Adjust for vehicle ownership	0.19	0.53	0.51	0.04
Assume all commute VKT still made	0.10	0.43	0.13	0.02
Offset to maximum savings	36.5%	36.5%	61.1%	36.5%
Offset with all commute VKT still made	64.1%	48.8%	90.4%	63.2%

achieved. For the odd/even day driving ban, total fuel savings are about 34% when adjustments for household vehicle ownership are made and are about 14% when it is also assumed that all commute VKT are still made. These figures drop to 7% and 2% fuel savings for a one-day-in-ten driving ban.

Consensus estimate of reduction from driving ban policies

Driving ban policies are potentially quite effective in reducing fuel consumption, even given the offsetting adjustments shown in the calculations above. In assessing the most likely effectiveness of these types of policies, we believe that considering the offsets is critical even when travellers are responding to price increases and calls for altruism. Therefore, our best estimate assumes that all commute VKT will still be made. While the actual

Table 2-26

Estimate of fuel savings for odd/even day driving ban

	Japan/ RK	IEA Europe	US/ Canada	Australia/ NZ	Total
Million litres saved/day					
50% VKT reduction applied to all VKT	162	437	913	41	1 553
Adjust for vehicle ownership	127	341	559	32	1 059
Assume all commute VKT still made	82	317	233	17	649
Thousand barrels saved/day					
50% VKT reduction applied to all VKT	1 021	2 749	5 741	255	9 766
Adjust for vehicle ownership	797	2 146	3 516	199	6 659
Assume all commute VKT still made	516	1 992	1 467	109	4 083
Road transport fuel saved (%)					
50% VKT reduction applied to all VKT	48.6%	48.7%	48.6%	48.3%	49.1%
Adjust for vehicle ownership	38.0%	38.0%	29.8%	37.7%	33.5%
Assume all commute VKT still made	24.5%	35.3%	12.4%	20.7%	20.5%
Total fuel saved (%)					
50% VKT reduction applied to all VKT	27.2%	31.0%	37.2%	34.3%	33.9%
Adjust for vehicle ownership	21.2%	24.2%	22.8%	26.8%	23.1%
Assume all commute VKT still made	13.7%	22.4%	9.5%	14.7%	14.2%

offsets may represent other sources of increased driving, this provides a good rough assumption of the potential offsets that could occur. Table 2-28 presents "consensus" estimates for both the odd/even and 1-in-10-day driving ban policies. This assumes adequate levels of enforcement to effectively penalise those who disregard the ban.

Vehicle speed-reduction policies

Vehicle speeds and fuel consumption

One factor associated with fuel consumption is driver behaviour, in terms of the selection of speed and acceleration style. In general, excessive acceleration and speed will tend to increase fuel consumption (see section on

Table 2-27

Estimate of fuel savings for one-day-in-ten driving ban

	Japan/ RK	IEA Europe	US/ Canada	Australia/ NZ	Total
Million litres saved/day					
50% VKT reduction applied to all VKT	32	87	183	8	310
Adjust for vehicle ownership	20	56	71	5	152
Assume all commute VKT still made	12	45	18	3	78
Thousand barrels saved/day					
50% VKT reduction applied to all VKT	204	550	1 148	51	1 953
Adjust for vehicle ownership	130	349	447	32	958
Assume all commute VKT still made	73	284	110	19	486
Road transport fuel saved (%)					
50% VKT reduction applied to all VKT	9.7%	9.7%	9.7%	9.7%	9.8%
Adjust for vehicle ownership	6.2%	6.2%	3.8%	6.1%	4.8%
Assume all commute VKT still made	3.5%	5.0%	0.9%	3.6%	2.4%
Total fuel saved (%)					
50% VKT reduction applied to all VKT	5.4%	6.2%	7.4%	6.9%	6.8%
Adjust for vehicle ownership	3.5%	3.9%	2.9%	4.4%	3.3%
Assume all commute VKT still made	2.0%	3.2%	0.7%	2.5%	1.7%

Table 2-28

Consensus estimates of fuel savings from odd/even day and one-day-in-ten driving ban policies

	Japan/ RK	IEA Europe	US/ Canada	Australia/ NZ	Total
Odd/even day driving ban policy					
Thousand barrels saved/day	516	1 992	1 467	109	4 083
Road transport fuel saved (%)	24.5%	35.3%	12.4%	20.7%	20.5%
Total fuel saved (%)	13.7%	22.4%	9.5%	14.7%	14.2%
1 day in 10 driving ban policy					
Thousand barrels saved/day	73	284	110	19	486
Road transport fuel saved (%)	3.5%	5.0%	0.9%	3.6%	2.4%
Total fuel saved (%)	2.0%	3.2%	0.7%	2.5%	1.7%

"ecodriving" later in the chapter). However, the conditions under which these occur may be different. For example, high speeds may occur on motorways with unrestricted flow while slower speeds with hard accelerations may be more likely when traffic is highly congested.

Most estimates of fuel efficiency with respect to vehicle speeds are based upon average driving cycles, thereby incorporating some level of accelerations and a variety of different speeds into estimates. Table 2-29 (based on ORNL, 2003) shows average fuel efficiency related to average speeds. As can be seen, the best fuel efficiency is achieved at speeds between 30 and 60 mph, deteriorating at higher speeds. The lower efficiency numbers at lower speeds would tend to be biased by being based on driving conditions with more stop-and-go driving and more accelerations. However, these figures would provide a preliminary basis for estimating the effect of speed reductions on motorways to speed limits of 90 kph or 55 mph.

Table 2-30 (also from ORNL, 2003) shows the variation in fuel economy ratings derived from different driving test cycles used to calculate fuel economy levels. These vary somewhat for Japan, Europe and the United States. The United States driving cycles tend to have higher maximum acceleration than those for Europe and Japan, while the European cycle has a higher maximum speed level. Table 2-31 shows the variation between the driving cycles. In any case, this shows the difficulty of measuring average fleet fuel economy levels. At the high end, however, we suspect that percentage reductions would be fairly similar across different countries and thus the figures in Table 2-29 could form a basis for estimates of the effect of speed reduction policies.

A study by DIW (1996) found that a reduction in German motorway speeds to 100 km/h and to 80 km/h on other extra-urban roads could save 4.8% of fuel consumption from private travel. Table A-17 in the appendix provides estimates of speeding in European Union countries which indicate opportunities to reduce average speeds without any changes in legislation. Overall estimates suggest that between 30% and 60% of vehicles exceed posted speed limits (although we would be primarily concerned with speeding in excess of 60 mph).

Table 2-29

Vehicle speed impacts on fuel economy

Speed		Fuel economy averages	
Miles per hour	Kilometres per hour	Miles per gallon	Litres per 100 km
15	24.2	24.4	9.6
20	32.3	27.9	8.4
25	40.3	30.5	7.7
30	48.4	31.7	7.4
35	56.5	31.2	7.5
40	64.5	31	7.6
45	72.6	31.6	7.4
50	80.6	32.4	7.3
55	88.7	32.4	7.3
60	96.8	31.4	7.5
65	104.8	29.2	8
70	112.9	26.8	8.8
75	121.0	24.8	9.5

Impact of speed reduction (to/from)		Percentage change in fuel economy	
55-65 mph	88.7-104.8 kph	11.0%	–8.8%
65-75 mph	104.8-121.0 kph	17.7%	–15.8%
55-75 mph	88.7-121.0 mph	30.6%	–23.2%

Source: Oak Ridge National Laboratory (2003). Based on model years 1988–1997 automobiles and light trucks, based on tests of 9 vehicles.

Table 2-30

Fuel economy estimates for different driving cycles

Driving cycle	Projected fuel economy for a 1995 composite midsize vehicle	
	Litres per 100 km	Miles per gallon
Japanese 10/15 mode test cycle	13.4	17.5
New European Driving Cycle (NEDC)	10.7	22.0
US EPA city cycle (LA4)	11.9	19.8
US EPA highway cycle	7.3	32.1
US Corporate Average Fuel Economy cycle	9.8	23.9

Note: The 1995 composite midsize vehicle is an average of a Chevrolet Lumina, using the National Renewable Energy Laboratory's Advanced Vehicle Simulator (ADVISOR) model.
Source: ORNL, 2003.

Table 2-31

Comparison of US, European and Japanese driving cycles

	Time (seconds)	Percent of time stopped or decelerating	Distance (miles)	Average speed (mph)	Maximum speed (mph)	Maximum acceleration (mph/s)
Japanese 10/15 mode test cycle	631	52.3	2.6	14.8	43.5	1.78
New European Driving Cycle (NEDC)	1 181	24.9	6.8	20.9	74.6	2.4
US EPA city cycle (LA4)	1 372	43.2	7.5	19.5	56.7	3.3
US EPA highway cycle	765	9.3	17.8	48.2	59.9	3.3
US Corporate Average Fuel Economy cycle	2 137	27.9	10.3	29.9	59.9	3.3

Note: When comparing data between countries, one must realise that different countries have different testing cycles to determine fuel economy and emissions. This table compares various statistics on the European, Japanese and US testing cycles [for fuel economy measurements, the US uses the formula, 1/fuel economy = (0.55/city fuel economy) + (0.45/highway fuel economy)]. Most vehicles will achieve higher fuel economy on the US test cycle than on the European or Japanese cycles.

Speed reduction policies can consist of many different measures. Informational policies might be effective, especially for reducing excessive accelerations. Changes in maximum speed limits can also be highly effective during crises, as demonstrated by the success of this policy in the United States during the 1970s, at least upon initial implementation. As the United States experience demonstrates, maintaining enforcement is critical. The European Union is urging member states to implement more speed control policies, primarily by the introduction of speed cameras. These have been estimated to be highly effective, with average speed reductions (for all road categories) of about 7% (ICF Consulting/Imperial College, 2003). If the focus is on reducing motorway speeds, one can roughly estimate the average speed reductions by the number of speed cameras per km placed on the entire motorway network (data on network length is available). Rough estimates of speed reductions can then be estimated if the goal is to reduce speeds to, say, 55 mph during an emergency.

Speed reduction analysis

Estimates were made of the likely fuel consumption savings due to reduced maximum speeds on motorways. Because of the different effects of varying driving cycles, the analysis was simplified to model only the effects of a change in maximum steady state speeds (an impact factor was then applied to account for this policy affecting only a portion of motorway fuel consumption). These estimates were made for all individual IEA member countries based upon country-specific data and various averages assumed for different countries.

The basic methodology involved multiple steps to best estimate the total vehicular traffic and fuel consumption that would be subject to the policy:

■ Total road transport fuel consumption was obtained from IEA for all member countries. Data on population, motorisation registered vehicles (by vehicle class), total road vehicle-kilometres travelled (VKT, by vehicle class, where available), motorway-vehicle travelled (by vehicle class, where available), road tonne-kilometres moved, and road network (by road type) were gathered from IRTAD, Eurostat, various national statistical agencies and other sources.

■ Approximately one-half of the data points for registered vehicles were unavailable or problematic. In particular, data regarding the number of goods vehicles were problematic with data for heavy goods vehicles, light goods vehicles and light-duty passenger "trucks" (*e.g.* sport utility vehicles, pick-up trucks and vans) intermingled and characterised inconsistently or erroneously. These were estimated or adjusted by interpolation and extrapolation from the available data for kilometres, tonne-kilometre data and VKT data.

■ Where data were unavailable, an estimate of average annual VKT per vehicle obtained from other countries in the same IEA region was utilised to generate these estimates. For example, for Europe, the computed figures were approximately 13 800 km annually for light passenger vehicles, 41 000 km for buses, 29 000 for light goods vehicles and 84 000 for heavy goods vehicles.

■ Motorway VKT data were estimated next; here, few original data were available disaggregated by vehicle class. National level motorway data were first estimated where not available, based on the length of motorways and motorways' share of the primary and total road network. Total motorway VKT was then distributed across vehicle classes. On the basis of the limited available data and professional judgment, motorway VKT was distributed to each vehicle class proportional to its share of total VKT, except that heavy goods vehicles were weighted at double their total VKT share and light goods vehicles were weighted at three-quarters their total VKT share.

- Fuel efficiencies by country and vehicle class were then estimated where the data were lacking. While national totals are commonly available, the more disaggregate data have limited availability. Interpolation and estimation using VKT by vehicle class and available data were then made, cross-checking that fuel consumption estimated in this manner matched IEA data for total road fuel consumption in the country.

- Fuel consumption by vehicle class for all roadways and for motorways were then calculated by multiplying VKT by fuel efficiencies. For motorways, fuel consumption rates were increased by 10% to reflect the higher rates observed at motorway speeds compared to other roads.

- Fuel consumption savings from speed reductions were estimated using a commonly used fuel consumption equation standard to mechanical engineering texts (Delucchi *et al.*, 2000; Gillespie, 1992; Thomas and Ross, 1997; Ross, 1997; Mendler, 1993).

The variables considered were:

MPG = Miles per gallon
V = Velocity of the vehicle, in metres/second
CdA = Coefficient of drag (air resistance)
A = Frontal area of vehicle (square metres)
AD = Air density (1 184 kg/m^3)
CdR = Coefficient of drag (rolling resistance)
W = Gross vehicle weight (kg)
EE = Engine efficiency (percentage)
FE = Fuel energy (Btu/gallon)

These factors are combined in an equation such that:

$$MPG = FE / (Total\ resistance \bullet 2\ 546.7 \bullet V \bullet 0.44704 / EE)$$

Where total resistance = aerodynamic drag + rolling resistance:

$$(0.5 \bullet CdA \bullet AD \bullet FE \bullet V^3 / 745.7) + (W \bullet 9.81 \bullet CdR \bullet V / 745.7)$$

Table 2-32 lists the assumptions made for various vehicle types.

The above equation was then used to estimate the difference in fuel consumption at different steady state speeds for a given vehicle class. The vehicle characteristics utilised and fuel economy results of a steady state 50 mph travel pattern are provided in Table 2-32, while Table 2-33 shows the generalised effect of two different policies for reduced motorway maximum speeds and Table 2-34 provides the results based on each region's maximum speed limit.

Because of the driving cycle issues discussed earlier, a policy impact factor of 50% was then applied to the results of the equation when calculating the overall fuel consumption savings from each country, with results provided in Table 2-33. This impact factor was based on professional judgment of the likely overall effectiveness of reducing the maximum legal speed limit, accounting for the following factors:

■ Many vehicles are already travelling below the maximum speed limit; in some cases this may be due to driver preference or, in others, to congestion. For example, approximately one-third of tractor-trailer

Table 2-32

Vehicle characteristics and illustrative results of fuel consumption equation

	North America light duty passenger	Rest-of-world light duty passenger	Light goods vehicles	Heavy goods vehicles
Coefficient of drag, air (CdA)	0.42	0.38	0.50	0.60
Frontal area (m²)	2.4	1.9	4.0	4.5
Coefficient of drag, rolling resistance (CdR)	0.015	0.015	0.015	0.015
Speed (mph)	50	50	50	50
Speed (m/s)	22.4	22.4	22.4	22.4
Gross vehicle weight (lbs)	4 025	3 400	10 000	40 000
Gross vehicle weight (kg)	1 826	1 542	4 536	18 144
Engine efficiency	15%	15%	22%	26%
Miles per gallon	20.0	25.7	13.2	5.6
Litres per 100 km	11.8	9.2	17.9	41.7

Table 2-33

Results of steady state speed reduction

Percentage reduction in fuel consumption at steady speed reduction of 20 kph						
Reduced speed	**70 kph**	**80 kph**	**90 kph**	**95 kph**	**100 kph**	**110 kph**
Original speed	**90 kph**	**100 kph**	**110 kph**	**115 kph**	**120 kph**	**130 kph**
US passenger car	22.9	22.7	22.3	22.0	21.7	21.1
ROW light-duty passenger	21.4	21.3	21.1	20.9	20.7	20.2
Light goods vehicle	20.8	20.8	20.6	20.5	20.3	19.8
Heavy goods vehicle	10.8	11.4	11.9	12.0	12.2	12.5
Percentage reduction in fuel consumption at steady speed reduction to 90 kph						
Reduced speed	**90 kph**	**90 kph**	**90 kph**	**90 kph**	**90 kph**	**90 kph**
Original speed	**90 kph**	**100 kph**	**110 kph**	**115 kph**	**120 kph**	**130 kph**
US passenger car	0.0	12.0	22.3	26.9	31.1	38.7
ROW light-duty passenger	0.0	11.3	21.1	25.5	29.6	37.0
Light goods vehicle	0.0	11.0	20.6	25.0	29.0	36.4
Heavy goods vehicle	0.0	6.0	11.9	14.7	17.5	22.8

kph: kilometres per hour.

motorway VKT and three-fifths of all other vehicles' motorway VKT in the United States are on urban motorways rather than rural motorways.

- In the EU, heavy trucks are already regulated with speed governors and limited to 90 kph. We therefore assume no reduction in heavy truck speeds in the EU (although anecdotal evidence of higher truck speeds on some European motorways suggests that compliance is not always 100% and therefore some fuel savings from heavy trucks might occur from lowered speed limits and/or tighter enforcement).

- While the analysis uses the maximum posted motorway speed limit in each country (with the exception of Germany, where the highest recommended speed of 130 kph is used), some motorways may have lower posted speeds which may not be affected as much by the speed reduction policies.

- Some vehicles may not slow to the full extent of the policy's intended reduction. For example, many vehicles normally travelling 110 kph in a posted 110-kph zone may slow to only 95 or 100 kph if the limit is lowered to 90 kph.

Table 2-34

Estimate of fuel savings for speed reductions

	20 kph reduction	Reduction to 90 kph
	Litres saved/day (million)	
Japan/RK	3.7	2.4
IEA Europe	20.3	29.3
North America	53.0	57.8
Aus/NZ	0.9	1.1
Total	**78.0**	**90.6**
	Barrels saved/day (thousand)	
Japan/RK	23.6	15.0
IEA Europe	127.9	184.2
North America	333.5	363.4
Aus/NZ	5.7	7.2
Total	**490.7**	**569.7**
	Percent road transport fuel saved	
Japan/RK	1.2%	0.7%
IEA Europe	2.3%	3.3%
North America	2.9%	3.2%
Aus/NZ	1.1%	1.4%
Total	**2.5%**	**2.9%**
	Percent total fuel saved	
Japan/RK	0.6%	0.4%
IEA Europe	1.4%	2.1%
North America	2.2%	2.4%
Aus/NZ	0.8%	1.0%
Total	**1.7%**	**2.0%**

- A significant portion of motorway fuel consumption is influenced by acceleration patterns and other elements of the driving cycle. For example, adjustments to driving speed to account for merging, other vehicles travelling at different speeds, etc. will occur regardless of the maximum speed. The motorway fuel consumption attributable to these driving activities would not experience the same percentage reduction as does the motorway fuel consumption attributable to travel at a steady state speed.

Given these assumptions, the results in Table 2-33 are adjusted by 50%. Table A-18 in the appendix shows the estimated percentage reductions of total transport and total fuels for each IEA country. Regional totals are shown in Table 2-34. As can be seen, speed reduction measures appear most effective in Europe and North America potentially leading to about a 5% reduction in total road transport fuel use (or about 3% to 4% of total fuel use).

Consensus estimate of reduction from speed limit policies

Reductions in speeds during a fuel emergency can be implemented in many ways. For example, in the United States, during the 1970s fuel emergency, a national speed limit of 55 mph (90 km/h) was implemented. Initially, this policy was very effective, primarily because of altruistic behaviour and a determined enforcement regime. The British fuel emergency of 2000 suggested that free-flow speeds on motorways decreased with no change in policy, presumably from attempts to conserve fuel by individual drivers. This suggests that actual shortages can induce some beneficial behavioural responses even without enforcement of new speed limits.

Therefore, our consensus estimate assumes that information is provided to encourage drivers not to exceed 90 km/h. Supplementing this with an enforcement regime, whether through speed cameras or increased presence of traffic police, should be very effective. Table 2-35 provides a summary of the results of our best estimate which assumes a change in the legal speed limit and a comprehensive enforcement regime.

Table 2-35

Consensus estimate of effect of reducing speed limit to 90 km/hr

	Japan/ Korea	IEA Europe	US/ Canada	Australia/ NZ	Total
Thousand barrels saved per day	15	184	363	7	570
Percent road transport fuel saved	0.7%	3.3%	3.2%	1.4%	2.9%
Percent total fuel saved	0.4%	2.1%	2.4%	1.0%	2.0%

Promotion of "ecodriving"

The concept of ecodriving has been extensively researched in the Netherlands and is actively supported by a number of countries (*e.g.* Switzerland and Germany co-operate on the website "www.ecodrive.org"). Ecodriving is estimated to provide a potential reduction in fuel use of 10% to 20% for those who fully adopt it. Across an entire car-driving population, the reductions will be lower since not everyone will be reached or be willing to change behaviour. However, in an emergency, assuming a good level of awareness and co-operation from the general public, an average reduction in fuel consumption per kilometre of 5% appears possible. During the British fuel crisis there was some evidence that drivers drove more slowly to conserve fuel, so clearly during crisis conditions drivers are willing to adopt these types of strategies (Eves *et al.*, 2002). Tyre pressure adjustments alone (discussed below) can provide half of this benefit.

There are several basic rules to follow:

- Shift up as soon as possible
- Maintain a steady speed
- Anticipate traffic flow
- Decelerate smoothly
- Switch off the engine at short stops
- Reduce vehicle weight to the extent possible (*e.g.* remove unnecessary items from trunk)
- Avoid reducing vehicle aerodynamics from items attached to the exterior of the car
- Keep tyres properly inflated and purchase low-rolling resistance replacement tyres
- Use low-viscosity motor oils.

Ecodriving training also includes encouraging drivers to make full use of in-car devices such as fuel economy computers, shift indicator lights and cruise control. These can help indicate the efficiency of driving and/or improve it.

There are different ways to train individuals in ecodriving. The most effective is on-road training, but there are also training possibilities with simulators and Internet simulators. In order to encourage a very rapid shift in driving style, an aggressive advertising campaign that provides necessary details can be used to great effect.

Driving style and fuel consumption

Estimating fuel consumption reductions from improved driving styles (such as reductions in hard accelerations and hard breaking) is somewhat difficult, partly because it appears to be quite variable. ECMT/IEA (2005) reports on a California Air Resources Board study that compared the standard United States driving cycle (the Federal Test Procedure or FTP) against a more aggressive driving cycle (Unified Cycle or UC). The UC had average accelerations about 30% greater than the FTP city cycle, with maximum accelerations and decelerations of over 100% greater. For 17 cars tested, the UC led to between a 5% and 14% increase in fuel consumption. Further, it was found that more aggressive driving leads to greater fuel consumption when the horsepower/weight (HP/WT) ratio is smaller. A typical family sedan (HP/WT of 0.04) was found to have a 6% increase in fuel consumption for the more aggressive driving cycle.

The ECMT and IEA (2005) report on the differences between the FTP (city) and a more aggressive driving cycle on the basis of an analysis of motorway drivers done by the United States Environmental Protection Agency. They found that the more aggressive driving cycle led to a 25% to 48% increase in fuel consumption. They concluded that the average car would experience about a 33% fuel penalty, while more powerful cars would have about a 28% fuel penalty when driven more aggressively. The two driving cycles measured had about the same average speed, so this result is clearly due to changes in maximum speed and acceleration/deceleration behaviour.

Various technologies are available that provide drivers with feedback on the fuel consumption associated with their driving style and/or provide information on more efficient driving styles. For example, shift indicator lights provide feedback on the most efficient gear to drive in for manual transmission cars. ECMT/IEA (2005) reports fuel savings between 5% and 15% from proper gear shifting. Fuel economy indicators, such as computers, are also becoming increasingly popular in vehicles, although it is less clear whether drivers respond to the information that this provides (at least without training). Cruise control systems can result in 20% to 30% increases in fuel efficiency when used by aggressive drivers; however, there is undoubtedly some self-selection in choosing to use technologies that control aggressive behaviour.

Potential benefits of tyre pressure improvements

Maintaining the proper tyre pressure can have a significant effect on total fuel consumption. ECMT/IEA (2005) reports estimates of a 2.5% to 3.0% increase in fuel consumption for every pound per square inch (psi) below the optimal tyre pressure. CEC (2003) reports a somewhat lower estimate of a 1% increase per 1.0 psi below the optimal level. Table 2-36, reproduced from ECMT/IEA (2005) shows that a significant fraction of cars have their tyres under-inflated, suggesting some room for increased efficiency.

Development of tyres with lower rolling resistance could potentially lead to some improvements in fuel economy. CEC (2003) estimated that conversion to lower rolling-resistance tyres could lead to about a 3% reduction in fuel usage. Decreases in fuel consumption are greater under high-speed highway conditions. Simulations reported in CEC (2003) suggest that a 10% decrease in rolling resistance results in a 2% decrease in fuel consumption for highway conditions. For urban driving, the decrease is about 1% for a 10% decrease in rolling resistance. While rolling resistance tends to vary by tyre, research conducted in Germany suggests that 50% improvements in tyre rolling resistance are easily achievable over the next few years. Car manufacturers typically seek to have low rolling resistance tyres on new vehicles (mainly to comply with US fuel economy standards or European voluntary measures), which has acted as an incentive for the tyre industry to develop these more efficient tyres. However, replacement tyres typically are not marketed or bought for their fuel efficiency.

This suggests a possible policy option of either providing information to consumers or mandating specific rolling resistance standards for replacement

Table 2-36

Percentage under-inflation of tyres based on survey

Vehicle type	No. of tyres under-inflated by > 8 psi				
	0	1	2	3	4
Passenger cars	73	14	7	3	3
Pickups, SUVs and vans	68	13	10	4	6

Source: ECMT/IEA (2005). psi: pounds per square inch.

tyres. One impediment to establishment of these types of policies is that currently there are no standardised test procedures for measuring the fuel efficiency associated with tyres, but this could be done[4]. Another policy approach is to have tyre excise tax rates set so that consumers have incentives to purchase those with lower rolling resistance. In any case, promoting changes to purchase patterns of replacement tyres represents a longer-term effort – not something that makes sense to link to emergency plans.

In the short term, the only tyre-related policy likely to be implemented would involve educational and communication campaigns for drivers to maintain the maximum approved tyre pressure. Estimates were made of the likely fuel consumption savings due to reduced rolling resistance of fully inflated tyres. Our methodology used to analyse the effectiveness of such a policy was very similar to that used to evaluate speed reductions, involving the same sets of equations shown above. Again, because of the different effects of varying driving cycles, the analysis was simplified to model only the effects of a change in rolling resistance due to tyre pressure at various steady state speeds.

On the basis of the survey data (see Table A-19 in the appendix), it was estimated that the average light-duty vehicle tyre is under-inflated by three psi. This was estimated by cross-multiplying the percentages shown in Table 2-37 by an assumed under-inflation of 12 psi (for those tyres noted in the table as at least 8 psi under-inflated) and assuming half of all other tyres are under-inflated by an average of three pounds. The approximate midpoint was then chosen between the 2.8 calculated for passenger cars and 3.3 calculated for light-duty trucks. For heavy-duty vehicles, the studies cited above suggested approximately an average 5 psi shortfall in tyre pressure and 0.6% change in fuel economy; these values were directly adopted. For all vehicles, it was assumed that the policy could not be 100% effective due to mis-inflation, leakage and similar factors; thus tyres were estimated to remain an average of 1 psi under-inflated under the programme. Table 2-38 provides results of the fuel economy simulation of the tyre pressure change.

---

4. *California recently passed legislation that mandates state agencies to purchase more efficient replacement tyres and as part of this requires the development of consistent standards to measure their efficiency. See for example http://www.nrdc.org/media/pressreleases/031002.asp for a description.*

Table 2-37

Estimated benefits from tyre inflation campaigns

Reduction in fuel consumption at steady speed from improved tyre inflation							
Vehicle speed (kph)		**30**	**50**	**70**	**90**	**110**	**130**
Light-duty average under-inflation (psi)	Before campaign	3	3	3	3	3	3
	During campaign	1	1	1	1	1	1
Heavy-duty average under-inflation (psi)	Before campaign	6	6	6	6	6	6
	During campaign	1	1	1	1	1	1
Percentage reduction in fuel consumption							
US passenger car		– 3.6%	– 2.9%	– 2.2%	– 1.7%	– 1.3%	– 1.0%
Other country passenger car		– 3.7%	– 3.0%	– 2.4%	– 1.9%	– 1.5%	– 1.2%
Light goods vehicle		– 3.7%	– 3.1%	– 2.5%	– 1.9%	– 1.5%	– 1.2%
Heavy goods vehicle		-0.7%	-0.7%	-0.6%	-0.6%	-0.5%	-0.4%
Absolute reduction in fuel consumption (litres per 100 km)							
US passenger car		0.22	0.22	0.22	0.22	0.22	0.22
ROW passenger vehicle		0.19	0.19	0.19	0.19	0.19	0.19
Light goods vehicle		0.38	0.38	0.38	0.38	0.38	0.38
Heavy goods vehicle		0.24	0.24	0.24	0.24	0.24	0.24

Specification of motor oil grades

Different engine oil grades tend to result in different levels of fuel economy. Studies have generally found about a 1% to 2% increase in fuel efficiency when lower-viscosity oil is used in place of those grades most commonly used (ECMT/IEA, 2005). Efficiency improvements may be even larger during cold temperatures. This suggests one possible policy of requiring the use of low-viscosity oils in those cars where engine damage would not occur (probably nearly all cars, except for high-performance vehicles) or taxation policies to reduce the relative cost of low-viscosity oils. Estimates of relative efficiency are presented in Table 2-39 based on results reported in the ECMT/IEA report.

To adequately estimate the improvements in fleet fuel efficiency, one would also need to know which oils are currently in use. Sales data suggest that higher-viscosity 10W-30/40 oils are still the most frequently bought oils for oil changes while newer vehicles are normally filled with lower-viscosity oils at the factory (mainly 5W-30). Therefore, it might be possible to develop rough estimates of total fuel savings based on assumptions about current motor oil usage.

Table 2-38

Estimates of fuel savings from tyre inflation campaigns

Million litres saved/day	
Japan/RK	7.2
IEA Europe	19.9
North America	30.8
Aus/NZ	1.4
Total	**59.3**
Thousand barrels saved/day	
Japan/RK	45.5
IEA Europe	125.2
North America	193.9
Aus/NZ	8.6
Total	**373.2**
Road transport fuel saved (%)	
Japan/RK	2.3
IEA Europe	2.2
US/Canada	1.6
Australia/NZ	1.6
Total	**1.9**
Total fuel saved (%)	
Japan/RK	1.2
IEA Europe	1.4
US/Canada	1.3
Australia/NZ	1.2
Total	**1.3**

Table 2-39

Fuel efficiency improvements from low-viscosity oils

Lower-viscosity oil	10W-30	10W-40	5W-30
5W-30	1.2-2.0%	1.2-2.0%	–
5W-20	–	–	1.0-3.5%
0W-20	–	–	1.0-2.0%

Changing the ability of vehicle fleets to use lower-viscosity oils would be difficult to do in the short term. However, it could be included in a general programme for promoting ecodriving, with some potential benefits, especially over a fairly long emergency period (*e.g.* several months). Potential savings from efficient motor oils are included in our "consensus" estimates from ecodriving, below.

Consensus estimate of reductions from comprehensive ecodriving programme (including driving style, tyre inflation and improved motor oil use)

Policies to promote ecodriving would mainly rely upon an aggressive campaign of public information and education. Some technology aspects could also be promoted, such as on-board information systems for fuel economy and diagnostic systems for tyre pressure. However, this would take time to be implemented within the full fleet of vehicles and would not be effective in short-term emergency. The assumptions stated in the above analysis appear reasonable as representing the case of a good public education programme aimed at increasing tyre pressures. Our consensus estimate is based upon this and some additional benefits from other aspects of an ecodriving programme. Results are summarised in Table 2-40.

Table 2-40

Consensus estimates for comprehensive ecodriving campaigns

	Japan/ RK	IEA Europe	US/ Canada	Australia/ NZ	Total
Fuel saved per day (million litres)	16.2	43.7	91.3	4.1	155.2
Fuel saved per day (thousand barrels)	102.1	274.9	574.1	25.5	976.6
Road transport fuel saved (%)	5.0%	5.0%	5.0%	5.0%	5.0%
Total fuel saved (%)	2.7%	3.1%	3.7%	3.4%	3.4%

Alternative fuels

Another potential package of policy options would focus on shifting to other fuels. These would, for the most part, consist of long-term policies to diversify the fuel supply and would not be activated in short-term emergency conditions. Various alternative fuels are making small inroads into the transport fleet, most notably compressed natural gas (CNG) and liquid petroleum gas (LPG), used mainly in business fleets, and ethanol, being used primarily as a fuel additive in blends up to 5% or 10% in various IEA countries. Brazil has long led the world in ethanol-fuelled vehicles, with most cars running on blends with about 25% ethanol content (IEA, 2004).

A more recent development is production of vehicles able to use any blend ratio of gasoline and ethanol. These are known as flexible-fuel vehicles. Several models are now being marketed in the United States and Brazil. In countries with many flexible-fuel vehicles and wide availability of both gasoline and ethanol, it becomes possible for consumers to switch quickly between fuels. Government initiatives to urge switching on short notice, such as during emergencies, may also become viable. However, there must be plenty of ethanol stock available to handle the increased demand on short notice. Brazil appears close to achieving this situation, with wide availability of both gasoline and ethanol and a rapidly growing stock of flexible-fuel vehicles. No IEA countries appear close to achieving similar conditions. For these reasons, we do not evaluate these policies further.

Chapter summary

A variety of different approaches to saving oil in hurry has been presented in this chapter. A summary of estimates, focused on the "consensus" estimates for each type of policy and sub-policy category, is presented in Table 2-41.

Clearly a wide variety of impacts can result from different types of policies and different formulations of similar policies. The total oil savings across IEA countries ranges from about ten thousand barrels per day for some transit options to four million for an odd/even day driving ban. While there is considerable uncertainty in the estimates, they provide a guide for which policies are likely to provide small, medium, or large reductions, as summarised above in the executive summary.

Table 2-42 shows the variation by region for each measure, expressed in barrels of oil saved as well as percentage reductions. There is a wide variation in results by region for nearly all measures, and there are substantial differences in the relative effectiveness of different measures in different

Table 2-41

Summary of effects of policies across all IEA countries

	Policy context to achieve savings	Thousand barrels per day saved	Road transport fuel savings (%)	Total fuel savings (%)
Policies to increase public transit usage	50% reduction in current public transit fares	280	1.4%	1.0%
	Free public transit	563	2.8%	2.0%
	Increase off-peak public transit service	188	0.9%	0.7%
	Increase peak and off-peak public transit service	232	1.2%	0.8%
	Allow existing bus and car-pool lanes to operate 24 hours	17	0.1%	0.1%
	Add additional lanes for buses with 24-hour usage	34	0.2%	0.1%
Policies to increase car-pooling	Build car-pool lanes along all motorways, add park-and-ride lots, comprehensive programmes to match riders	1 240	6.2%	4.3%
	Small programme to match riders, public information	170	0.9%	0.6%
Increasing telecommuting	Public information to employers on benefits of telecommuting, minor investment to facilitate	730	3.7%	2.6%
Compressed 4/40 work week	Public information to employers on benefits of compressed work weeks	520	2.6%	1.8%
Driving bans and signage	Odd/even driving ban. Provide police enforcement, appropriate information	4 100	21%	14%
	1 day in 10 driving ban. Provide police enforcement, appropriate information and signage	490	2.4%	1.7%
Speed limit reduction	Reduce speeds to 90kph. Provide police enforcement or speed cameras, appropriate information and signage	570	2.9%	2.0%
Ecodriving campaign	Provide public information and other support and incentives for ecodriving	370	5.0%	3.5%

Note: actual road transport fuel consumption in IEA countries in 2001: 20 088 thousand barrels/day; total petroleum consumption: 28 813 thousand barrels/day.

Table 2-42

Estimated fuel savings for each IEA region

	Japan/RK	IEA Europe	US/Canada	Aus/NZ	Japan/RK	IEA Europe	US/Canada	Aus/NZ	Japan/RK	IEA Europe	US/Canada	Aus/NZ
Current road transport fuel consumption (2001), thousand bbls per day	2 101	5 643	11 816	528								
Current total petroleum fuel consumption (2001), thousand bbls per day	3 760	8 882	15 428	743								
Public transit:	**Fuel savings (thousand bbls per day)**				**Percent of road transport fuel saved**				**Percent of total petroleum fuel saved**			
50% fare reduction	64	172	42	3	3.1%	3.0%	0.4%	0.5%	1.7%	1.9%	0.3%	0.3%
100% fare reduction	128	344	85	6	6.1%	6.1%	0.7%	1.2%	3.4%	3.9%	0.6%	0.8%
Off-peak service	59	96	31	3	2.8%	1.7%	0.3%	0.5%	1.6%	1.1%	0.2%	0.3%
Peak and off-peak service	74	117	38	3	3.5%	2.1%	0.3%	0.5%	2.0%	1.3%	0.3%	0.3%
Bus and HOV enhancement	3	11	4	0.2	0.12%	0.19%	0.03%	0.05%	0.07%	0.12%	0.02%	0.03%
Bus and HOV expansion	5	21	7	0.5	0.24%	0.38%	0.06%	0.09%	0.14%	0.24%	0.05%	0.07%
Car-pooling infrastructure and programme	125	277	800	38	6.0%	4.9%	6.8%	7.2%	3.3%	3.1%	5.2%	5.1%
Car-pooling programme	13	41	112	6	0.6%	0.7%	1.0%	1.1%	0.3%	0.5%	0.7%	0.8%
Telecommuting	88	102	523	21	4.2%	1.8%	4.4%	4.0%	2.3%	1.2%	3.4%	2.9%
Compressed four-day work week	61	71	367	15	2.9%	1.3%	3.1%	2.8%	1.6%	0.8%	2.4%	2.0%
Odd/even day driving ban	516	1 992	1 467	109	24.5%	35.3%	12.4%	20.7%	13.7%	22.4%	9.5%	14.7%
One day in ten driving ban	73	284	110	19	3.5%	5.0%	0.9%	3.6%	2.0%	3.2%	0.7%	2.5%
Speed limits at 90 kph	15	184	363	7	0.7%	3.3%	3.2%	1.4%	0.4%	2.1%	2.4%	1.0%
Eco-driving campaign	102	275	574	26	5.0%	5.0%	5.0%	5.0%	2.7%	3.1%	3.7%	3.4%

regions. In general, transit-oriented policies work best in the regions where transit ridership is already very important: Europe and Japan/Korea. In contrast, car-pooling measures are relatively more effective in those regions with the highest driving shares: United States/Canada and Australia/New Zealand.

The potential of telecommuting and flexible work policies also is least effective in the European region, relative to other regions. This is due to relatively lower current levels of solo car driving for commute trips to work. Thus, the benefit of a telecommuting or flexible work schedule policy is relatively greater in those countries that currently have more solo car commute trips.

On the other hand, driving bans appear most effective in Europe and least effective in North America. This is a function of the relative levels of household car ownership in each region. Average car ownership per household is highest in North America, which means that households are more likely to have at least one car available on any given day that a driving ban is enforced (as these are usually set by licence plate number).

Speed limit reduction and enforcement policies appear most effective in Europe and North America. This is due to both relatively higher motorway usage (relative to Japan/RK and Australia/NZ) and (in the case of Europe) higher maximum speed limits, providing more benefit from a reduction. Another fuel economy-related measure, ecodriving campaigns, is estimated to give similar levels of effectiveness across regions.

The following chapter focuses on estimating the costs of these different policies. As will be seen, the cost per barrel of oil saved also varies tremendously across policy types, but not necessarily in a way that correlates with the oil savings of each policy. Very few provide large reductions at low cost per unit reductions.

3. IMPLEMENTATION COST AND COST-EFFECTIVENESS OF VARIOUS POLICY OPTIONS

While the policies and strategies analysed in Chapter 2 show a large range of potential effectiveness in reducing fuel consumption, their relative cost-effectiveness is another important criterion for determining which policies might be implemented. Cost-effectiveness can be defined as the net cost, in terms of policy implementation, needed to save a barrel of fuel. However, many other benefits besides fuel savings can also be attributable to these policies. For example, congestion reduction can lead to major cost savings in users' travel time; reduced pollutant emissions improve health and the environment locally and globally; and reductions in crashes and injuries (for example, from speed limit reduction) provide major benefits. These three factors are usually the key external costs associated with transport. Even more difficult to measure are reductions or increases in accessibility, mobility and mode choice (and the value of these changes). All of these are beyond the scope of this study. Instead, the focus here is on relatively easily quantified financial and "implementation" costs of the measures, most of which are borne by governments. These estimates are then compared to the fuel reductions achieved, in order to derive an estimate of the "implementation cost per barrel saved". This is a useful, though clearly incomplete, guideline for governments to use when choosing between policy options.

General considerations

Our estimates of the relative implementation cost and cost-effectiveness of the various policies are based upon an assumption that the average duration of an emergency is 90 days. Therefore, we consider the total barrels saved over this time frame (assuming the same average savings per day). Since we compare the implementation cost of the policies to the total barrels of oil saved, their ratio is the cost per barrel saved. Measures can be considered cost-effective if their cost per barrel saved is less than the cost of a barrel – which during an emergency could be fairly high, perhaps well above $50 per barrel.

Cost-effectiveness for each measure was calculated as the cost both per litre and per barrel saved. As petroleum savings were represented in daily terms, costs were similarly converted to a daily rate. For marginal costs, this was a simple conversion of time units. For one-off or fixed costs, they were divided by 90 (days) to show their cost-effectiveness during the emergency[5].

Some of the policies evaluated require only a public information campaign to make them effective. We assume the same costs for these campaigns for each country and for each policy. These are shown in Table 3-1. Public announcement costs are based on drafting fact sheets, transmitting information to government officials, disseminating information via e-mail or "broadcast" faxes, disseminating press releases, providing copy for radio and television public service announcements and other activities. Costs for preparing information pamphlets are assumed to be $0.02 per employed person in the country (*e.g.* $1 million for a country with 50 million employed people). We assume staff costs at $100 000 per annum, or slightly less than $25 000 prorated over the 90 days of the emergency. We assume that governments can obtain free access to most media, but we assume miscellaneous costs of $15 000 related to delivering public announcements.

Table 3-1

Assumed costs of a public information campaign by region

	Thousand US dollars				
	Japan/ RK	IEA Europe	US/ Canada	Australia/ NZ	Total
Pamphlet preparation and printing	1 700	2 656	2 891	168	7 415
Staff costs	49	419	49	49	567
Public announcement costs	30	255	30	30	345
Total	1 779	3 330	2 970	247	8 328

Other potential costs are considered in more detail and in some cases a range of potential costs is provided, leading to a range of cost-effectiveness

---

5. *Ninety days was chosen to represent a typical supply emergency length. For one-off costs such as outreach campaigns (e.g. tyre pressure awareness), these figures may be conservative with regard to overall cost-effectiveness as there is likely some longer-term educational and fuel savings benefit. For the infrastructure investment costs (e.g. bus and car-pool lanes), these figures significantly understate the likely long-term cost-effectiveness of the measure by including only their benefits during the emergency period.*

measures. Specific assumptions for each policy are detailed below, followed by a summary of the relative effectiveness of each.

Cost-effectiveness estimates by policy type

Cost-effectiveness of public transit strategies

Costs assumed for public transit policies are shown in Table 3-2. Costs for the two fare reduction measures (50% and 100% reduction) were calculated by taking the average fare per existing public transit trip (from the Millennium Database; UITP, 2001), and multiplying it by the number of existing public transit trips in the region and by 50% or 100% to calculate the revenue forgone. This approach assumed no net additional cost for the new public transit trips. In reality, there are confounding additional marginal costs (possible need for additional service provision, security) and benefits (reduced labour costs from reduced/no fare collection/enforcement; reduced dwell times due to reduced payments, etc.). It is also important to distinguish that, in fact, only these additional "confounding" costs represent true economic costs – since they represent additional resource requirements (and they represent economic benefits when fewer resources are needed). The loss of revenues from fare reductions are not economic costs – instead they represent wealth transfers. In this case there is no change in the activity – providing transit service – only a change in who pays for it (the government, or taxpayers, rather than transit riders). Thus by focusing on lost fare revenues, we are measuring revenue impacts to the government, not true economic costs. But these implementation "costs" to the government are an important consideration when choosing among measures to cope with oil supply disruptions.

Table 3-2

Public transit cost data

	Japan/ RK	IEA Europe	US/ Canada	Australia/ NZ
Average fare revenue per public transit trip	$1.22	$0.64	$0.65	$0.97
Average operating cost per vehicle-km	$4.26	$5.02	$4.21	$3.20
Cost to extend bus/HOV lane hours ($/day/km)	$4.26	$5.02	$4.21	$3.20
Cost to stripe additional bus/HOV lane-km ($/km/90 days)	$137.59	$138.35	$137.54	$136.53

HOV: high-occupancy vehicle.

For the two measures related to increasing transit level of service (one with increases during off-peak times, the other with increases during both peak and off-peak times), costs were obtained by estimating the additional peak and off-peak vehicle-kilometres of service provided and multiplying by the average operating cost per kilometre (both from the Millennium Database). The ratio of peak to off-peak vehicle-km was estimated as 0.4 to 0.6, based on UITP statistics (UITP, 1997).

The third set of transit measures involves designating special lanes for buses and increasing the use of existing bus lanes to 24 hours. As for the impact estimates, costs for these measures were estimated by assuming that for each kilometre of separated bus lanes, there are 10 km of bus/car-pool shared priority lanes and 100 km of total bus routes. For cost estimation, this ratio is not terribly important, since although the level of benefit changes with different ratios, the level of cost changes proportionately and cost-effectiveness changes very little. Enforcement costs for the measure were derived from speed enforcement calculations by ICF Consulting and Imperial College London (2003) as approximately $5 per lane-kilometre daily[6].

For conversion of existing lanes to new bus priority lanes, costs were estimated by first calculating the total additional bus priority lane-kilometres per region at two linear metres per 1 000 urban residents. This was multiplied by $12 000 per kilometre for road painting and signage, which was estimated as four times more expensive as striping and signing bicycle lanes (typically $3 per linear metre). This cost was divided by the expected 90-day duration of the supply emergency.

Cost-effectiveness estimates for these various strategies are shown in Table 3-3. As can be seen, all are quite costly per barrel of oil saved. Adding bus lane infrastructure appears to be the most cost-effective of these strategies and would be of moderate cost-effectiveness if implemented without increasing the operating period of existing bus lanes. Extending the operating period of bus lanes loses much of its cost-effectiveness because the enforcement costs are being applied to the relatively small target audience of off-peak bus ridership.

---

6. *This estimate is based on a marginal cost for supplemental enforcement of $200 per person-day and assuming that each enforcement agent could cover 40 lane-km.*

Table 3-3

Public transit policy cost-effectiveness

Measure	Cost per barrel of oil saved				
	Japan/ RK	IEA Europe	US/ Canada	Australia/ NZ	Average
Reduce public transit fares by 50%	$1 002	$507	$469	$969	$658
Reduce public transit fares by 100%	$1 002	$507	$469	$969	$658
Increase weekend and off-peak service to peak levels (increase frequency by 40%)	$906	$1 313	$1 222	$1 611	$1 171
Increase weekend and off-peak service and increase peak service frequency by 10%	$845	$1 225	$1 140	$1 504	$1 171
Convert all car-pool and bus lanes to 24-hour bus priority usage	$43	$79	$75	$129	$73
Convert all car-pool and bus lanes to 24-hour bus priority usage and implement an additional two linear metres of lanes per 1 000 urban residents	$31	$44	$50	$77	$43

Cost-effectiveness of car-pooling strategies

Our consensus estimates of the effectiveness of car-pooling strategies focused on two potential policies. One was focused on creating car-pool lanes on motorways while the other assumed a programme of education and encouragement on the benefits of car-pooling.

Car-pool lanes can be added by either physical construction of new lanes or restriping and adding signage in existing lanes. In our cost analysis we assume that adding new lanes has a cost of $2.5 million per km. Restriping costs are $12 000 per kilometre (the same as for bus lane restriping). We assume two cases: 1) where car-pool lane capacity is added for all motorways; and 2) where capacity is added only on urban motorways. For North America, where most regions either have car-pool lanes or already have them programmed for construction, we assume that only 50% of existing urban motorways would need additional car-pool lanes. For a programme of public information and education we assume costs as shown in Table 3-1.

Table 3-4 displays the cost-effectiveness results for the car-pooling strategies. As can be seen, any programme of actual construction of infrastructure is very expensive per barrel of oil saved. On the other hand, the much lower costs of

Table 3-4

Car-pool policy cost-effectiveness

Measure	Cost per barrel of oil saved				
	Japan/ RK	IEA Europe	US/ Canada	Australia/ NZ	Average
Construction of car-pool lanes along all motorways	$939	$4 974	$1 550	$755	$1 928
Restriping existing lane to car-pool along all motorways	$5	$25	$8	$4	$10
Construction of car-pool lanes along all urban motorways	$272	$1 443	$450	$219	$559
Restriping existing lane to car-pool along all urban motorways	$1.48	$7.82	$2.44	$1.19	$3.03
Provide information on car-pooling benefits	$1.56	$0.90	$0.30	$0.49	$0.54

restriping existing infrastructure, especially if limited to urban motorways, are relatively cost-effective. Public information and education campaigns are generally very cost-effective.

Cost-effectiveness of work-trip reduction policies

Telecommuting policies may be more feasible if employees are provided with computers and broadband access so that they can more easily work from home. This may not be needed for every employee as home computer ownership is quite high, especially among that segment of the population that has "telecommutable" jobs. We assume that 50% of employees would need computers purchased for them to enable telecommuting and that the cost is $1 500 per computer. As can be seen, at this level of equipment provision, this type of policy is not cost-effective.

Providing information to encourage people to telecommute, however, can be very cost-effective. (In this case, "information" includes developing a programme that companies participate in, where they commit to allowing certain employees to telecommute during emergencies.) The same applies to encouraging people (or employers) to adopt flexible work schedules. For the telecommuting policy we assume no difference in the effect with and without the purchase of computers. In reality, one might expect that providing computers would make the policy more effective, but as our numbers clearly show, even a doubling of effectiveness would still not make this cost-effective. Or alternatively, if a public information campaign were ten times less effective, the cost per barrel saved would be less than $2 or $3 per barrel.

For these cost calculations we make no assumptions regarding whether worker productivity would be either positively or adversely affected. This may vary for individual jobs but on average we would expect any adverse effects to be balanced with productivity-enhancing effects. Results are shown in Table 3-5.

Table 3-5

Work-trip reduction policy cost-effectiveness

	Cost per barrel of oil saved				
Measure	**Japan/ RK**	**IEA Europe**	**US/ Canada**	**Australia/ NZ**	**Average**
Telecommuting with 50% purchase of computers	$3 529	$4 744	$1 007	$1 462	$1 842
Telecommuting with public information campaign, company commitments	$0.17	$0.27	$0.05	$0.10	$0.09
Compressed work week with public information campaign, company commitments	$0.32	$0.52	$0.09	$0.18	$0.18

Cost-effectiveness of driving bans

The costs of implementing a policy of driving bans will consist primarily of providing information to the public and an adequate enforcement programme. We assume the same programme costs as indicated above. An additional cost would be putting adequate signage in place to inform people. We assume that one sign would be installed for every 5 km of motorway at $5 000 per sign. Policing costs are more substantial and may consist of overtime payments for existing police or traffic officers, or increases in policing staff. We assume this cost at one officer per 100 000 employed people[7] at a $200 000 annual rate per officer (prorated over 90 days).

---

7. *This low ratio of additional officers is because the vast majority of the enforcement will be conducted by officers already patrolling; this merely represents the marginal enforcement effort, and is balanced by the high overtime cost of providing this additional marginal enforcement.*

In the estimates shown in Table 3-6, we see that these policies are generally very cost-effective. We make no distinction in the effectiveness of the policy with and without police enforcement. Clearly, it could be much less effective without adequate enforcement, but even if it were ten times less effective, the policy is still very cost-effective. If our policing cost estimates are relatively low, these results clearly show that even a doubling of our estimate would make this a cost-effective policy. The more stringent odd/even day policy is also more cost-effective than a one-day-in-ten ban, as the costs are the same.

Table 3-6

Driving ban policy cost-effectiveness

Measure	Cost per barrel of oil saved			
	Japan/ RK	IEA Europe	US/ Canada	Australia/Average NZ
Odd/even day ban with police enforcement and signage	$1.08	$0.67	$1.22	$0.60 $0.92
Odd/even day ban with signage only	$0.22	$0.32	$0.70	$0.21 $0.44
One-in-ten day ban with police enforcement and signage	$7.67	$4.70	$16.23	$3.46 $7.71
One-in-ten day ban with signage only	$1.56	$2.27	$9.33	$1.18 $3.72

It should be recognised that these bans may have some additional costs in terms of reduced accessibility and mobility options (particularly for single-vehicle households with limited access to alternative modes, or multi-vehicle households whose vehicles' licence plates coincidentally end in the same number). Estimates of these costs would be pure conjecture.

Cost-effectiveness of speed reduction policies

The costs associated with reducing speeds are essentially related to the enforcement regime that is in place. While we would expect a public information campaign to lead to some driving adjustments through altruistic behaviour, especially over the short run, experience suggests that some enforcement is needed. Enforcement can be achieved either by increasing traffic policing or by speed cameras.

For policing costs, we assume that at least one additional traffic officer (or overtime equivalent) would be needed per 50 000 employed people at a cost of $200 000 per officer (prorated over 90 days). Speed camera costs are estimated at €21 000 (about $26 000) based on estimates from ICF Consulting and Imperial College London (2003) and we assume one speed camera is needed for every 10 km of motorway. We also include costs for increased signage based on figures estimated above.

As with other measures, there are additional costs and benefits to users (increased safety, increased travel time, lower fuel costs) that are not included here. In particular, past studies indicate that the value of safety benefits can be very large for a speed reduction policy.

Cost-effectiveness results are shown in Table 3-7. Surprisingly we find little difference in cost-effectiveness numbers between adding police versus speed cameras. Clearly, different assumptions on the costs of these measures would lead to different results. With the exception of Japan/RK, we generally find this policy to have a moderate level of cost-effectiveness.

Table 3-7

Speed reduction policy cost-effectiveness

Measure	Cost per barrel of oil saved				
	Japan/ RK	IEA Europe	US/ Canada	Australia/ NZ	Average
Speed limit – increased traffic police, signage	$68.36	$11.21	$7.10	$16.00	$10.14
Speed limit – speed cameras only, signage	$21.64	$11.40	$9.46	$9.70	$10.40

Cost-effectiveness of ecodriving information campaign

Costs for implementing an ecodriving information campaign are as estimated above for a public information campaign. If a 5% reduction in road transport fuel use can be achieved, as described for this measure in Chapter 2, then it would be extremely cost-effective. This is shown in Table 3-8.

Table 3-8

Ecodriving campaign cost-effectiveness

	Cost per barrel of oil saved				
Measure	**Japan/ RK**	**IEA Europe**	**US/ Canada**	**Australia/ NZ**	**Average**
Ecodriving programme	$0.19	$0.13	$0.06	$0.11	$0.09

Summary of cost-effectiveness results

The cost-effectiveness calculations shown above suggest some general conclusions about what type of policies can be implemented cheaply, relative to the reductions in fuel consumption they may deliver. However, the cost estimates here neglect many important (but hard-to-measure) factors and should not be taken as a complete cost-effectiveness analysis.

Policies that add substantial infrastructure or otherwise require a large budgetary outlay tend not to be cost-effective. In particular, the public transit policies of reducing fares or increasing service frequency are not cost-effective[8]. Construction of new car-pool lanes is also not a cost-effective policy for reducing fuel consumption. In addition, purchasing computers to enable telecommuting would not be effective and actually might not even be necessary to implement a telecommuting policy.

Among the infrastructure policies, restriping of motorway lanes to car-pooling lanes is moderately cost-effective. Public transit policies that increase bus lane usage are also moderately cost-effective in some cases, as are speed limit reduction policies. Their cost-effectiveness is primarily determined by the costs associated with enforcement. Other studies have found that reducing speeds is highly cost-effective for safety reasons regardless of the benefits of reducing fuel usage.

The most cost-effective policies are clearly those that can be implemented mainly with a public information campaign. This includes telecommuting and

---

8. *Increased public transit service may still make sense in terms of facilitating other measures, particularly driving bans, even if the direct fuel consumption reductions are not cost-effective.*

flexible work policy promotion as well as ecodriving campaigns. Even if the cost of an aggressive public information campaign were much higher than we have assumed in this analysis, the cost per barrel saved for these options may be quite low. Odd/even day driving restrictions also are very cost-effective despite some of the enforcement and signage costs. This is due mainly to the large potential savings that can be achieved by driving restriction policies. One-in-ten-day driving bans are less cost-effective since the costs are about the same as for an odd/even day ban. However, such travel-restriction measures may impose large costs on society in terms of lost mobility and economic activity.

In all cases, several potentially important types of costs are not accounted for here. These include the value of travel time, safety impacts and pollutant emissions impacts. Some policies may have large impacts in one or other of these areas. For example, speed limit reduction may have its biggest cost in terms of increased travel times and its biggest benefit in terms of reduced number and severity of accidents (and fatalities and injuries). A careful analysis of these types of impacts is suggested for countries making their own estimates.

4. CONCLUSIONS

This report has evaluated the ability to reduce short-term fuel consumption in the transport sector via the use of demand restraint policies, in the event of emergency supply constraints. The focus of this analysis has been to evaluate a range of policy options commonly used under normal circumstances by transport planners to manage transport demand, primarily to reduce traffic congestion and environmental impacts associated with transport. This analysis differs in that it views these same measures under the much different circumstances of a temporary supply disruption or sudden severe price spike.

This analysis is based, to the extent possible, upon existing estimates within the literature and experience from past fuel crises. In some cases, given the shortage of data covering emergency situations, expert judgment has been used to estimate behaviour and response to policies in such situations. The transport literature generally analyses the longer-term effects associated with various policies under normal fuel supply conditions; we have tried to estimate likely effects under conditions of supply disruption, including the altruistic effects that would influence travel behaviour under emergency conditions.

The basic approach has been to evaluate the impact of a variety of measures if applied individually during an emergency, given the necessary emergency planning and preparation before an emergency occurs. In most cases the measures have the effect of reducing light-duty vehicle travel, either by reducing demand or encouraging shifting to public transit or other modes. We have evaluated the following general approaches:

- Increases in public transit usage

- Increases in car-pooling

- Telecommuting and working at home

- Changes in work schedules

- Driving bans and restrictions

- Speed limit reductions

- Ecodriving campaign

An important conclusion is that measures that rely on altruistic behaviour and provide information to consumers may be able to provide substantial reductions in fuel consumption at very low cost. This includes car-pooling, flexible work schedules, telecommuting and ecodriving campaigns. Conversely, those policies that involve large upfront investments, especially in terms of adding infrastructure, are not very cost-effective for reducing fuel consumption (although there may be other reasons to implement them).

Those policies that are more restrictive appear capable of providing large reductions in fuel consumption. For example, odd/even day driving bans give the largest estimated reductions in fuel consumption. However, such restrictive policies may be unpopular and may impose substantial "hidden" costs on society in terms of lost mobility.

The results presented here provide fairly rough estimates of effectiveness. We have used real data disaggregated to individual countries, where possible. The main source of uncertainty in our results is the extent to which people will respond to the measures in emergency situations, especially those measures based on providing information. We expect that in most cases, people will naturally seek alternative transport options during a fuel emergency due to both the increase in the price of fuel and actual supply constraints. In such a situation, most people will welcome, and respond to, additional information and mobility options provided by governments. Previous short-term emergency conditions suggest that some altruistic behaviour will occur, whether prodded by price or actual concern about helping society weather an emergency.

Another important consideration is the synergistic effect of adopting a mix of policies. For example, driving restrictions will be more feasible to implement if public transit options have been increased or if telecommuting is actively promoted. We did not evaluate the interactions between policies, but in many cases we would expect adoption of a broad package of policies to provide the greatest reductions in fuel consumption.

The cost-effectiveness calculations presented here are incomplete and are meant to providing an indication regarding how implementation costs compare to expected fuel savings. Many important costs and benefits of policies, such as those related to safety and mobility, are not included in our estimates. However, the estimates provide some important indications: *e.g.* policies that require major investments tend not to be cost-effective while those that are relatively easy to implement are more cost-effective. Further research is needed to better estimate full costs associated with policies and more closely link these to detailed policy specifications. Individual countries may be best suited to conduct such analysis.

APPENDIX: DATA SOURCES AND CALCULATIONS FOR ESTIMATES IN THIS REPORT

Various data sources have been used to estimate fuel consumption reductions from the implementation of transport demand management policies. An attempt has been made, wherever possible, to use country-specific data which are then aggregated to the four IEA regions. These regions are North America (United States and Canada), Europe (the European Union plus Norway and Switzerland), Australia/New Zealand and Japan/Republic of Korea.

We have strived to be consistent in the sources of data for the various analyses conducted. Consistent data are needed for our estimates of current fuel consumption, current vehicle-kilometres of travel (VKT), fuel intensity and vehicle occupancy rates. This section briefly notes the sources and methods used to calculate these data, with special attention paid to any omissions. In addition, we used the Millennium Cities Database for some of our analysis (UITP, 2001). Since these data are only based on a sampling of urban areas, total country and regional estimates were obtained by applying figures calculated to the entire region using current population data.

Fuel consumption

Base fuel consumption levels are as supplied by the IEA. These include total fuel consumption, gasoline and diesel consumed by the transport sector, and gasoline and diesel consumed by the road transport sector. Original values were supplied as metric tons of fuel. These were converted to barrels using country-specific conversion factors as supplied by the IEA. Total consumption for each country is listed in Table A-1 in units of thousand barrels. Countries included in these calculations are shown. Note that Europe includes only the European Union-15 plus Switzerland and Norway. Totals for each region are shown in Table A-2.

Table A-1

Total fuel consumption for each country, 2001 data
(thousand barrels per year)

	Total fuel consumption (all sectors)	Total transport fuel consumption	Total road transport fuel consumption
Australia	230 750	162 687	156 920
Austria	70 706	45 859	45 449
Belgium	111 669	60 486	59 614
Canada	454 800	324 373	304 841
Denmark	48 392	28 070	26 846
Finland	50 203	29 952	28 893
France	521 850	333 314	325 758
Germany	740 905	430 913	424 856
Greece	87 606	45 198	42 326
Ireland	47 654	26 989	26 691
Italy	384 081	276 307	273 532
Japan	1 087 454	608 505	595 412
Rep. of Korea	284 983	158 326	149 993
Luxembourg	18 029	12 410	12 320
Netherlands	110 098	78 686	76 068
New Zealand	40 456	29 986	29 232
Norway	46 627	29 027	23 750
Portugal	61 370	43 706	43 027
Spain	307 278	219 626	207 175
Sweden	76 805	52 971	51 956
Switzerland	92 123	39 139	38 981
United Kingdom	466 651	306 986	299 362
United States	5 176 245	3 988 461	3 885 971

Table A-2

Total petroleum fuel consumption for each region and total for all regions, 2001 data

	Japan/ RK	IEA Europe	US/ Canada	Aus/NZ	Total
All sectors, billion litres/year	218.2	515.4	895.2	43.1	1 671.8
All sectors, thousand bbls/day	3 760	8 882	15 ,428	743	28 813
Transport sector, billion litres/year	121.9	327.4	685.6	30.6	1 165.6
Transport sector, thousand bbls/day	2 101	5 643	11 816	528	20 088
Road transport sector, billion litres/year	118.5	319.0	666.2	29.6	1 133.3
Road transport sector, thousand bbls/day	2 042	5 498	11 482	510	19 531

One discrepancy was found in these data. New Zealand did not have any consumption of diesel for the road transport sector listed. However, there was a relatively large amount of diesel consumption listed under "transport-non-specified". This was included as being from road transport in our calculations.

Vehicle-kilometres travelled

Data on vehicle-kilometres travelled (VKT) was obtained from the International Road and Traffic Accident Database (IRTAD) supplied by the IEA. For the most part we used 2001 data to be consistent with our fuel consumption data. For Australia and New Zealand, only year 2000 data were available. The Netherlands also only had 2000 data. VKT data for the United Kingdom do not include Northern Ireland, which is a relatively small fraction of the total VKT. Table A-3 shows the details. VKT includes estimates for all forms of motorised road transport.

Table A-3

Vehicle travel estimates from IRTAD

Country	Year	Billion VKT
Australia	2000	18.5
Austria	2001	75.5
Belgium	2001	91.5
Canada	2001	310.2
Denmark	2001	46.7
Finland	2001	46.7
France	2001	551.0
Germany	2001	620.3
United Kingdom	2001	473.9
Ireland	2001	37.8
Japan	2001	790.8
Netherlands	2000	126.7
New Zealand	2000	37.2
Norway	2001	33.3
Republic of Korea	2001	27.4
Switzerland	2001	59.8
United States	2001	44.8

Fuel intensity calculations

Fuel intensity calculations for each region were based on total VKT and road transport fuel consumption for each region. Two minor caveats would be the slight underestimation of VKT for the United Kingdom (from the omission of Northern Ireland) and the use of 2000 VKT data for the Netherlands. A rough calculation suggests that the error this introduces is less than 1%. Fuel intensity for Australia and New Zealand is based on 2000 figures for VKT and 2001 fuel consumption figures. This might result in estimating a slightly less efficient fleet for this region, but this would again be within the margin of error for estimates of this type. In addition, VKT data were not available for Greece, Portugal, Luxembourg, Spain and Italy, and thus these were excluded from the fuel intensity calculations. Final values used in our calculations are shown in Table A-4.

Table A-4

Average fuel intensity for each region

Region	Litres/100 km
Japan/RK	11.13
IEA Europe	10.17
United States/Canada	13.91
Australia/NZ	13.34

The Millennium Cities Database

The Millennium Cities Database (UITP, 2001) contains detailed transport statistics for a large sampling of urban areas throughout the world. Those cities within the IEA countries with complete data were used in our analysis and are listed in Table A-5.

Table A-5

IEA cities in the Millennium Database

City	Country	City	Country
Amsterdam	Netherlands	Milan	Italy
Athens	Greece	Montreal	Canada
Atlanta	United States	Munich	Germany
Barcelona	Spain	Nantes	France
Berlin	Germany	New York	United States
Berne	Switzerland	Newcastle	United Kingdom
Bologna	Italy	Osaka	Japan
Brisbane	Australia	Oslo	Norway
Brussels	Belgium	Ottawa	Canada
Calgary	Canada	Paris	France
Chicago	United States	Perth	Australia
Copenhagen	Denmark	Phoenix	United States
Denver	United States	Prague	Czech Republic
Düsseldorf	Germany	Rome	Italy
Frankfurt	Germany	Ruhr (region)	Germany
Geneva	Switzerland	San Diego	United States
Glasgow	United Kingdom	San Francisco	United States
Graz	Austria	Sapporo	Japan
Hamburg	Germany	Seoul	Republic of Korea
Helsinki	Finland	Stockholm	Sweden
Houston	United States	Stuttgart	Germany
Lille	France	Sydney	Australia
London	United Kingdom	Tokyo	Japan
Los Angeles	United States	Toronto	Canada
Lyon	France	Vancouver	Canada
Madrid	Spain	Vienna	Austria
Manchester	United Kingdom	Washington	United States
Marseille	France	Wellington	New Zealand
Melbourne	Australia	Zurich	Switzerland

To normalise estimates based upon this sampling of urban areas we used data on total population for each region, total urban population for each region and the percentage of total urban population represented by our Millennium Database sample. These data are shown in Table A-6 with the calculated percentages used for normalisation to represent the entire region. These data were used for those policies expected to be applied just in urbanised areas. Note that total European Union population was used (omitting Switzerland and Norway).

Table A-6

Population normalisation factors applied to sample Millennium cities

	Japan/ RK	IEA Europe	US/ Canada	Australia/ NZ
Region population total (millions)	174.9	388.6	319.8	23.4
Population in urban areas (millions)	139.3	308.0	246.9	19.8
Urban population, percentage	79.6%	79.3%	77.2%	84.9%
Population for cities in Millennium Database (millions)	71.5	70.6	68.8	10
Total urban population, percentage	51.3%	22.9%	27.9%	50.3%

Table A-7

Data on Millennium cities' public transit use, 1997

	Units	Japan/ RK	IEA Europe	US/ Canada	Australia/ NZ
Number of cities in database		4	37	14	5
Population in metropolitan area of covered cities	million persons	71.5	77.7	68.8	10.0
Daily public transit trips per capita	trips/person	0.76	0.59	0.16	0.19
Daily private transport trips per capita	trips/person	1.1	1.4	3.0	3.1
Overall average trip distance	km	10.1	7.9	11.9	8.7
Overall average trip distance by car	km	12.2	12.4	13.2	9.9
Overall average trip distance by public transit	km	14.2	7.8	10.5	12.9
Annual car travel – vehicle km per capita	thousand kilometres	2.9	4.5	10.7	7.4
Annual car travel – passenger-km per capita	thousand kilometres	4.4	6.1	15.0	11.4
Annual public transit boardings per capita	boardings	413	330	87	84
Bus boardings per capita	boardings	102	147	57	40
Minibus boardings per capita	boardings	–	1.2	0.3	–
Tram boardings per capita	boardings	5	112	9	36
Light-rail boardings per capita	boardings	4	27	11	–
Metro boardings per capita	boardings	107	108	37	–
Suburban rail boardings per capita	boardings	201	37	3	35
Heavy-rail boardings per capita	boardings	308	146	43	–
Annual public transit passenger-km per capita	passenger-kilometres	4 046	1 668	634	918
Bus passenger km per capita	passenger-kilometres	685	633	335	293
Minibus passenger km per capita	passenger-kilometres	–	2	6	–
Tram passenger km per capita	passenger-kilometres	13	255	48	194
Light-rail passenger km per capita	passenger-kilometres	24	123	83	–
Metro passenger km per capita	passenger-kilometres	644	562	308	–
Suburban rail passenger km per capita	passenger-kilometres	2 703	512	119	578
Heavy-rail passenger km per capita	passenger-kilometres	3 347	1 017	453	–
Public transit average seat occupancy (load factor)	persons/seat	0.93	0.47	0.33	0.27
Bus seat occupancy	persons/seat	0.63	0.47	0.30	0.28

Table A-7 (continued)

Data on Millennium cities' public transit use, 1997

	Units	Japan/ RK	IEA Europe	US/ Canada	Australia/ NZ
Minibus seat occupancy	persons/seat	–	0.22	0.61	–
Tram seat occupancy	persons/seat	0.87	0.57	0.70	0.54
Light-rail seat occupancy	persons/seat	0.75	0.54	0.46	–
Metro seat occupancy	persons/seat	1.10	0.75	0.40	–
Suburban rail seat occupancy	persons/seat	0.92	0.32	0.33	0.27
Heavy-rail seat occupancy	persons/seat	1.02	0.46	0.40	–
Public transit operating cost per vehicle-km	USD/vkt	4.36	5.02	4.21	3.20
Public transit operating cost per passenger-km	USD/passenger-km	0.11	0.27	0.29	0.19
Private passenger transport energy use per capita	MJ/person	10 690	15 335	50 862	29 610
Public transit energy use per capita	MJ/person	1 187	1 136	889	795
Total transport energy use per capita	MJ/person	11 876	16 371	51 751	30 405
Energy use per private passenger vehicle-km	MJ/km	3.2	3.3	4.7	3.9
Energy use per public transit vehicle-km	MJ/km	13.5	14.5	24.9	14.9
Energy use per bus vehicle-km	MJ/km	16.2	16.2	27.2	17.0
Energy use per minibus vehicle-km	MJ/km	–	8.8	8.4	–
Energy use per tram wagon-km	MJ/km	10.3	13.3	15.6	10.1
Energy use per light rail wagon-km	MJ/km	9.5	19.5	16.8	–
Energy use per metro wagon-km	MJ/km	11.2	11.5	20.4	–
Energy use per suburban rail wagon-km	MJ/km	11.0	14.8	49.5	12.1
Energy use per heavy rail wagon-km	MJ/km	11.7	12.9	25.0	–
Energy use per private passenger-km	MJ/pkm	2.2	2.5	3.4	2.6
Energy use per public transit passenger-km	MJ/pkm	0.4	0.8	1.8	0.9
Energy use per bus passenger-km	MJ/pkm	0.9	1.1	2.4	1.7
Energy use per minibus passenger-km	MJ/pkm	–	2.5	1.3	–
Energy use per tram passenger-km	MJ/pkm	0.5	0.7	0.7	0.4
Energy use per light-rail passenger-km	MJ/pkm	0.3	0.8	0.6	–
Energy use per metro passenger-km	MJ/pkm	0.2	0.5	1.3	–
Energy use per suburban rail passenger-km	MJ/pkm	0.3	0.9	1.4	0.5
Energy use per heavy-rail passenger-km	MJ/pkm	0.2	0.5	0.9	
Overall energy use per passenger-km	MJ/pkm	1.4	2.1	3.3	2.4

Table A-8

Population normalisation factors applied to sample Millennium cities

	Units	Japan/ RK	IEA Europe	US/ Canada	Australia/ NZ
Number of cities in database		4	37	14	5
Population in database, metropolitan areas	persons	71 504 732	77 689 088	68 849 627	9 979 051
Daily public transit trips per capita	trips/person	0.76	0.59	0.16	0.19
Daily bus trips per capita	trips/person	0.20	0.39	0.11	0.13
Daily rail trips per capita	trips/person	0.56	0.20	0.06	0.06
Daily private transport trips per capita	trips/person	1.08	1.43	3.04	3.06
Daily public transit trips in database	trips	54 164 834	45 663 920	11 337 239	1 935 936
Daily private transport trips in database	trips	77 225 111	110 922 753	209 073 367	30 515 938
Daily bus and tram trips in database	trips	14 374 629	30 315 808	7 393 753	1 325 797
Daily metro and suburban rail trips in database	trips	39 790 205	15 348 111	3 943 485	610 139
Bus and tram mode share	Percentage	26.5%	66.4%	65.2%	68.5%
Region population total	persons	174 927 000	388 604 000	319 798 000	23 373 000
Population in urban areas	persons	139 311 535	308 028 328	246 852 788	19 839 138
Percentage urban	percentage	79.6%	79.3%	77.2%	84.9%
Metro pop. for cities in Millennium Database	persons	71 504 732	77 689 088	68 849 627	9 979 051
Percent of total urban population	percentage	51.3%	25.2%	27.9%	50.3%
Estimated public transit trips in region	daily trips	105 522 763	181 052 206	40 649 834	3 848 779
Estimated private transport trips in region	daily trips	150 448 296	439 796 002	749 635 595	60 667 869
Estimated peak public transit trips in region	daily trips	58 037 520	99 578 713	22 357 408	2 116 829
Estimated off-peak public transit trips in region	daily trips	47 485 244	81 473 493	18 292 425	1 731 951
Estimated bus trips in region	daily trips	28 004 343	120 198 704	26 510 409	2 635 778
Estimated peak bus trips in region	daily trips	15 402 389	66 109 287	14 580 725	1 449 678
Estimated off-peak bus trips in region	daily trips	12 601 954	54 089 417	11 929 684	1 186 100
Bus reserved route length	kilometres	49.2	319.0	215.5	14.6

Vehicle occupancy estimates

Estimates of average vehicle occupancy were calculated using the Millennium Database, which contains data on VKT and passenger-kilometres of travel (PKT) for each city. These were grouped by region and aggregate totals calculated. These represent urban vehicle occupancy levels and therefore may not be representative of rural areas. However, most of our policies for increasing vehicle occupancy would likely have their greatest impact in urban areas. Therefore, we use these numbers as shown in Table A-9. We also use our own judgment on vehicle occupancy for commuter trips, for which there were no reliable data at the level we sought. These are also shown in Table A-9 and would only apply to policies that influence commuter trips.

Table A-9

Estimates of average vehicle occupancy rates

	Japan/ RK	IEA Europe	US/ Canada	Australia/ NZ
Average urban vehicle occupancy	1.50	1.37	1.40	1.53
Average commute vehicle occupancy	1.25	1.15	1.10	1.10

Transit ridership analysis estimates

Based on the literature review conducted for the study, effectiveness factors and elasticities were selected for variants of each of the three transit measures discussed in Chapter 2 (fare reductions, service enhancements and lane prioritisation). Table A-10 shows how the effectiveness factors were applied and the resulting impacts on private vehicle trip reduction and daily fuel use, for each of the six measures. Table A-11 presents the summary results in terms of passenger car VKT reduced and litres of petroleum saved. Table 2-11 in the main text shows this for barrels per day saved and percentage reductions in fuel use.

Table A-10

Effectiveness of public transit measures: trips diverted from private vehicles

Measure	Impact	Million trips per day			
		Japan/ RK	IEA Europe	US/ Canada	Australia/ NZ
Reduce public transit fares by 50%	Increase in transit ridership (apply own price elasticity) (-0.4 Europe and Asia; -0.3 North America and Oceania)	21.1	36.2	6.1	0.6
	Reduction in trips in private vehicles*				
	• Apply 60% diversion factor to estimate private vehicle trips reduced	12.6	21.8	3.6	0.3
	• Apply cross-price elasticity (-0.10) to private transport trips	7.5	22.0	37.5	3.0
Reduce public transit fares by 100 %	Apply own-price elasticity (-0.4 Europe and Asia; -0.3 North America and Oceania) to public transit trips	42.2	72.4	12.2	1.2
	Reduction in trips in private vehicles*				
	• Apply 60% diversion factor to estimate private vehicle trips reduced	25.2	43.5	7.4	0.8
	• Apply cross-price elasticity (-0.10) to private transport trips	15.0	44.0	75.0	6.1
Increase weekend and off-peak service frequency by 40% (to peak levels)	Apply own-time out-of-vehicle elasticity (0.50) to off-peak public transit trips	11.6	19.9	4.5	0.4
	Apply 60% diversion factor to estimate private vehicle trips reduced	6.9	12.0	2.7	0.3
Increase off-peak service as above plus increase peak service frequency by 10%	Apply own-time out-of-vehicle elasticity (0.50) to peak and off-peak public transit trips	14.5	24.9	5.6	0.5
	Apply 60% diversion factor to estimate private vehicle trips reduced	8.7	14.9	3.3	0.3
Convert all HOV and bus lanes to 24-hour bus priority usage.	Apply own-time in-vehicle elasticity (0.4) to a 10% average time-saving on off-peak public transit trips	0.5	2.1	0.5	0.05
	Apply 60% diversion factor to estimate private vehicle trips reduced	0.3	1.4	0.3	0.03
Convert all HOV and bus lanes to 24-hour bus priority usage.	Apply own-time in-vehicle elasticity (0.4) to a 15% average time-saving on off-peak public transit trips and 5% for peak trips	1.1	4.6	1	0.01
	Apply 60% diversion factor to estimate private vehicle trips reduced	0.6	2.7	0.6	0.06

*Note: for reduction in trips in private vehicles, results of two methods are shown in two rows; only the lower estimate is used in subsequent calculations such as the following table.

Table A-11

Effectiveness of public transit measures: summary results

		Japan/ RK	IEA Europe	US/ Canada	Australia/ NZ
Assumptions used for all measures:					
• Average private vehicle trip distance (VKT)		12.20	12.36	13.20	9.90
• Average in-use LDV fuel consumption (L/100km)		11.13	10.17	13.91	13.34
Measure	**Impact**				
Reduce public transit fares by 50%	Private vehicle trips reduced (millions)	7.5	21.8	3.6	0.3
	Private VKT reduced (millions)	91.5	268.8	47.5	3.0
	Million litres saved	10.2	27.3	6.6	0.4
Reduce public transit fares by 100%	Private vehicle trips reduced (millions)	15	43.5	7.4	0.8
	Private VKT reduced (millions)	183.0	537.7	97.0	7.4
	Million litres saved	20.4	54.7	13.5	1.0
Increase weekend and off-peak service frequency by 40% (to peak levels)	Private vehicle trips reduced (millions)	6.9	12.0	2.7	0.3
	Private VKT reduced (millions)	84.2	148.3	35.6	3.0
	Million litres saved	9.4	15.1	5.0	0.4
Increase off-peak service as above plus increase peak service frequency by 10%	Private vehicle trips reduced (millions)	8.7	14.9	3.3	0.3
	Private VKT reduced (millions)	106.1	183.5	43.6	3.0
	Million litres saved	11.8	18.7	6.1	0.4
Convert all HOV and bus lanes to 24-hour bus priority usage	Private vehicle trips reduced (millions)	0.3	1.4	0.3	0.03
	Private VKT reduced (millions)	3.7	16.7	4.0	0.30
	Million litres saved	0.4	1.7	0.6	0.04
Convert all HOV and bus lanes to 24-hour bus priority usage	Private vehicle trips reduced (millions)	0.6	2.7	0.6	0.06
	Private VKT reduced (millions)	7.3	33.4	7.9	0.59
	Million litres saved	0.8	3.4	1.1	0.08

VKT: vehicle-kilometres travelled.
LDV : light-duty vehicle.

Car-pooling estimates

Table A-12 displays the coefficients used by McDonald and Noland (2001) which were collected from a variety of sources. These are derived from regional travel demand models estimated with multinomial logit choice models and provide some feel for the range of estimates that have been found in practice. Noland and McDonald also model trip time rescheduling in response to changes in congestion levels. This level of detail may not be needed when trying to model effects during a fuel shortage. The key coefficient values to consider are the travel time coefficient parameters which give an indication of how sensitive mode switching may be and any mode-specific parameters associated with HOV usage. The basic format of these models follows a random utility formulation implemented as a multinomial or nested logit model.

Table A-12

Nested logit model coefficients from McDonald and Noland (2001)

Model type	Variable	Value
Mode choice	Logsum for single-occupant vehicle (SOV)	0.684
	Logsum for high-occupancy vehicle (HOV)	0.224
	HOV delay coefficient	– 2.04
	HOV constant	– 2.0
Lane choice	Logsum for express lanes	0.1
	Logsum for mixed flow lanes	0.65
	Toll coefficient	– 0.532
	Lane constant	– 1.0
Time of day	Travel time coefficient	– 0.106 SOV – 0.045 HOV
	Coefficient for schedule delay-early (SDE)	– 0.065 SOV – 0.054 HOV
	Coefficient for schedule delay-late (SDL)	– 0.254 SOV – 0.362 HOV
	Coefficient for dummy variable for late arrival (DL)	– 0.58 SOV – 1.14 HOV

This can be estimated as the probability of choosing j conditional on the choice set i.

$$P(j\,|\,i) = \frac{e^{(Ui\, +\, \beta LS)}}{\sum_{k} e^{(Ui\, +\, \beta LS)}}$$

Where *Ui* represents the utility of each choice as a function of the parameter estimates. *LS* represents any log-sum coefficients if this is a nested logit form of the model. These methods are normally used in detailed travel demand models.

Table A-13

Car-pooling – impacts of adding one person to every car trip

	Japan/ RK	IEA Europe	US/ Canada	Australia/ NZ	Total
(Initial) average vehicle occupancy	1.50	1.37	1.40	1.53	
Daily urban VKT (millions) from Millennium sample of cities	529	830	1,964	203	3 526
Daily PKT (millions)	792	1,137	2,756	310	2 238
Daily VKT when adding one person to every car trip (millions)	318	479	1,148	123	2 068
VKT saved per day (millions)	211	350	817	80	1 458
Percentage VKT reduction	39.9%	42.2%	41.6%	39.4%	41.3%
Litres saved per day (millions)	24	36	114	11	185
Barrels saved per day (thousands)	148	224	715	67	1 154
Barrels saved per day, prorated for all urban areas (thousands)	289	977	2,560	134	3 960
Barrels saved per day, prorated for entire region (thousands)	363	1 233	3 320	158	5 073
Percentage saved urban areas	13.8%	17.3%	21.7%	25.4%	19.7%
Percentage of fuel used for transport saved, entire region	17.3%	21.9%	28.1%	30.0%	25.3%
Percentage of total fuel consumption saved, entire region	9.6%	13.9%	21.5%	21.3%	17.6%

Table A-14

Car-pooling – impacts of adding one person to every car trip on urban area motorways

	Japan/ RK	IEA Europe	US/ Canada	Australia/ NZ	Total
(Initial) average vehicle occupancy	1.50	1.37	1.40	1.53	
Percentage of total VKT on motorways	9.2%	22.5%	24.1%	24.1%	
Daily VKT on urban area motorways (millions)	49	186	474	49	759
Daily PKT on motorways (millions)	71	255	666	74	1 066
Daily motorway VKT when adding one person for trips on motorways (millions)	30	107	277	30	337
VKT saved per day (millions)	19	79	197	19	314
Litres saved per day (millions)	2.2	8.0	27.4	2.6	32
Barrels saved per day (thousands)	14	50	172	16	252
Barrels saved per day, prorated for all urban areas (thousands)	26	220	618	32	897
Barrels saved per day, prorated for entire region (thousands)	33	277	800	38	1 149
Percentage saved urban areas	1.3%	3.9%	5.2%	6.1%	4.5%
Percentage of fuel used for transport saved, entire region	1.6%	4.9%	6.8%	7.2%	5.7%
Percentage of total fuel consumption saved, entire region	0.9%	3.1%	5.2%	5.1%	4.0%

Table A-15

Car-pooling – impacts of adding one person to every commute trip

	Japan/ RK	IEA Europe	US/ Canada	Australia/ NZ	Total
(Initial) average vehicle occupancy, commute trips	1.3	1.2	1.1	1.1	
Daily VKT on commute trips (millions)	804	1 025	3 846	162	
Daily PKT on commute trips (millions)	1 006	1 179	4 230	178	6 593
Daily VKT when adding one person for all commute trips (millions)	447	548	2 014	85	3 094
VKT saved per day (millions)	358	477	1 831	77	2 743
Litres saved per day, entire region (millions)	40	48	255	10	353
Barrels saved per day, entire region (thousands)	250	305	1 603	65	2 223
Percentage of fuel used for transport saved, entire region	11.9%	5.4%	13.6%	12.3%	11.1%
Percentage of total fuel consumption saved, entire region	6.7%	3.4%	10.4%	8.7%	7.7%

Table A-16

Car-pooling – impacts of a 10% reduction in motorway VKT due to increased car-pooling

	Japan/ RK	IEA Europe	US/ Canada	Australia/ NZ	Total
Daily VKT on motorways (billions)	49	186	474	49	759
Daily motorway VKT with 10% reduction (billions)	44	167	427	44	682
VKT saved per day (millions)	5	19	47	5	74
Litres saved per day (thousands)	539	1 897	6 587	654	9 677
Barrels saved per day (thousands)	3	12	41	4	60
Barrels saved per day, prorated for all urban areas (thousands)	7	52	149	8	215
Barrels saved per day, prorated for entire region (thousands)	8	66	192	10	276
Percentage saved, urban areas	0.3%	0.9%	1.3%	1.6%	1.1%
Percentage of fuel used for transport saved, entire region	0.4%	1.2%	1.6%	1.8%	1.4%
Percentage of total fuel consumption saved, entire region	0.2%	0.7%	1.3%	1.3%	1.0%

Table A-17

European speed data

Country	Type of road	Speed limit	Vehicle types	Speeding statistics as compiled by each member State				Original data source
				Statistic 1	Statistic 2	Statistic 3	Statistic 4	
Austria	Motorway Rural main road Built-up area	130 100 50	Cars Cars Cars	Mean = 116 Mean = 90.5 Mean = 53.4	SD = 17.6 SD = 13.8 SD = 8	V85 = 134 V85 = 104 V85 = 61	Obs = 15,000 Obs = 24,000 Obs = 16,000	1996, FACTUM
Denmark	Rural road (avg May/Oct) Motor road (avg May/Oct) Motorway (avg May/Oct)	70/80 70/80 70/80/	Not listed Not listed Not listed	Mean = 112.1 Mean = 93.9 Mean = 88.6				Danish Road Directorate, 1995
Denmark	Single lane rural Motorway	89/90 100-130	Cars Cars	67% over limit 40% over limit				Danish Road Directorate, 1994
Finland	Rural (average winter/summer) Rural Motorways Motorways	80 100 100 120	All All All All	Mean = 82.5 Mean = 9 0 Mean = 98.4 Mean = 111.6	Over 80 kph = 66.1% Over 100 kph = 19.8% Over 100 kph = 49% Over 120 kph = 33.4%	Over 90 kph = 18.7% Over 110 kph = 4.2% Over 110 kph = 18.3% Over 130 kph = 10.7%	Over 100 kph = 3.7% Over 120 kph = 0.8% Over 120 kph = 4.2% Over 140 kph = 0.5%	1995, Finnish Road Admin.
Finland	Single lane rural Motorway Motorway	80/90 100/110 100-130	Cars Cars Cars	52% over limit 23% over limit 15% over limit				Mäkinen, 1990
France	Urban Single lane rural Motorway Motorway	50 80/90 100/110 100-130	Cars Cars Cars Cars	64% over limit 58% over limit 44% over limit 40% over limit				ONSR, 1994

Table A-17 (continued)

European speed data

Country	Type of road	Speed limit	Vehicle types	Statistic 1	Statistic 2	Statistic 3	Statistic 4	Original data source
				Speeding statistics as compiled by each member State				
Germany	Residential	30	Cars	74% over limit				Blanke, 1993
Ireland	Single lane rural	80/90	Cars	36% over limit				Crowley, 1991
	Two-lane rural	100	All	Mean = 85	SD = 12.5	V90 = 100	% speeding = 15	?
	Two-lane rural (avg of three)	80	All	Mean = 75	SD = 12.9	V90 = 89	% speeding = 28	
Netherlands	Motorways	100	Not listed	Mean = 104.1				1994, Project Bureau IVS
	Motorways	120	Not listed	Mean = 111.5				
Netherlands	Single lane rural	80/90	Cars	40% over limit				SVOV, 1994
	Motorway	100	Cars	55% over limit				
	Motorway	120	Cars	20% over limit				
Portugal	Two-lane rural	90	Cars	90 kph or more = 5.5%	95 kph or more = 2.9%	No cars ⩾ 110 km/h	Obs = 15 380	1996, TRANSPOR
	Residential in Catalonia	30/40	Cars	97-98% over limit				CdeC, 1992/1993
Spain	Urban	50	Cars	71% over limit				DGT, 1993
	Single lane rural	80/90	Cars	16% over limit				
	Motorway	100/110	Cars	22% over limit				
	Motorway	100-130	Cars	25% over limit				

Table A-17 (continued).

European speed data

Country	Type of road	Speed limit	Vehicle types	Speeding statistics as compiled by each member State				Original data source
				Statistic 1	**Statistic 2**	**Statistic 3**	**Statistic 4**	
Sweden	Not listed	30	All	30 kph or more = 76%	40 kph or more = 24%	50 kph or more = 6%		1996, Vägverket
	Not listed	50	All	50 kph or more = 58%	60 kph or more = 12%	70 kph or more = 2%		
	Rural	70	All	70 kph or more = 75%	80 kph or more = 40%	90 kph or more = 14%	100 kph or more = 2%	
	Rural	90	All	90 kph or more = 50%	100 kph or more = 17%	110 kph or more = 5%	120 kph or more = 1%	
	Rural	110	All	110 kph or more = 33%	120 kph or more = 11%	130 kph or more = 2%	140 kph or more = 1%	
	Motorway	90	All	90 kph or more = 80%	100 kph or more = 46%	110 kph or more = 17%	120 kph or more = 3%	
	Motorway	110	All	110 kph or more = 50%	120 kph or more = 22%	130 kph or more = 7%	140 kph or more = 1%	
United Kingdom	Urban	30 mph	Cars	Mean = 33	72% > limit	38% > 35 mph	Obs. = 2 515 000	1996, Transport Statistics GB
	Urban	30 mph	Trucks	Mean = 30	55% > limit	21% > 35 mph	Obs. = 101 000	
	Urban	40 mph	Cars	Mean = 37	28% > limit	10% > 45 mph	Obs. = 1 251 000	
	Urban	40 mph	Trucks	Mean = 33	14% > limit	3% > 45 mph	Obs. = 73 000	
	Single-lane rural	60 mph	Cars	Mean = 47	10% > limit	2% > 70 mph	Obs. = 13 156 000	
	Single-lane rural	40 mph	Trucks	Mean = 44	68% > limit	22% > 50 mph	Obs. = 2 125 000	
	Two-lane rural	70 mph	Cars	Mean = 68	47% > limit	11% > 80 mph	Obs. = 1 093 000	
	Two-lane rural	50 mph	Trucks	Mean = 55	85% > limit	12% > 60 mph	Obs. = 1 645 000	
	Motorways	70 mph	Cars	Mean = 70	55% > limit	18% > 80 mph	Obs. = 71 218 000	
	Motorways	60 mph	Trucks	Mean = 57	24% > limit	1% > 70 mph	Obs. = 18 724 000	

Table A-18

Percentage fuel consumption savings from reduction in steady state speed

	Motorway speed limit (kph)	% VKT on motorway	Speed reduced by 20 kph					Speed reduced to 90 kph			
			% fuel use on motorway	Light-duty passenger	Bus	Light goods	Heavy goods	Light-duty passenger	Bus	Light goods	Heavy goods
Australia	105.0	11%	15%	21%	21%	21%	12%	21%	21%	21%	21%
Austria	130.0	23%	26%	20%	20%	20%	0%	37%	37%	37%	0%
Belgium	120.0	34%	40%	21%	20%	20%	0%	30%	30%	30%	0%
Canada	110.0	25%	31%	22%	21%	21%	0%	21%	21%	21%	21%
Denmark	110.0	21%	23%	21%	21%	21%	0%	21%	21%	21%	0%
Finland	120.0	9%	11%	21%	20%	20%	0%	30%	30%	30%	0%
France	130.0	20%	23%	20%	20%	20%	0%	37%	37%	37%	0%
Germany	130.0	33%	37%	20%	20%	20%	0%	37%	37%	37%	0%
Greece	100.0	12%	14%	21%	21%	21%	0%	11%	11%	11%	0%
Ireland	112.7	3%	4%	21%	20%	20%	0%	25%	25%	25%	0%
Italy	130.0	15%	19%	20%	20%	20%	0%	37%	37%	37%	0%
Japan	100.0	9%	10%	21%	21%	21%	11%	11%	11%	11%	11%
Rep. of Korea	100.0	20%	25%	21%	21%	21%	11%	11%	11%	11%	11%
Luxembourg	120.0	22%	27%	21%	20%	20%	0%	30%	30%	30%	0%
Netherlands	120.0	45%	48%	21%	20%	20%	0%	30%	30%	30%	0%
New Zealand	100.0	8%	10%	21%	21%	21%	11%	11%	11%	11%	11%
Norway	90.0	2%	2%	21%	20%	20%	0%	0%	0%	0%	0%
Portugal	120.0	12%	15%	21%	20%	20%	0%	30%	30%	30%	0%
Spain	120.0	46%	49%	21%	20%	20%	0%	30%	30%	30%	0%
Sweden	110.0	14%	17%	21%	21%	21%	0%	21%	21%	21%	0%
Switzerland	120.0	34%	39%	21%	20%	20%	0%	30%	30%	30%	0%
United Kingdom	112.7	19%	21%	21%	20%	20%	0%	25%	25%	25%	0%
United States	104.6	23%	30%	23%	21%	21%	12%	21%	21%	21%	21%

Note: European Union regulations require speed governors on heavy goods vehicles, set to 90 km/hr. We have assumed no change in speeds for these vehicles. We assume the same applies to Norway and Switzerland.

Table A-19

Reductions in fuel consumption from tyre inflation campaign
(million litres)

	Light-duty passenger vehicle	Bus Europe	Light goods vehicle	Heavy goods vehicle	Total
Australia	273	7	57	71	408
Austria	107	2	62	6	177
Belgium	146	3	27	16	192
Canada	571	7	103	52	733
Czech Republic	61	3	21	7	92
Denmark	71	3	30	2	106
Finland	77	2	17	5	101
France	736	12	540	44	1 332
Germany	971	14	342	37	1 364
Greece	120	4	59	5	188
Hungary	51	3	27	7	88
Ireland	57	2	23	3	85
Italy	666	12	162	80	920
Japan	1 000	26	855	78	1 959
Rep. of Korea	309	7	272	91	679
Luxembourg	9	0	2	1	11
Netherlands	192	2	84	7	285
New Zealand	64	2	21	6	93
Norway	52	1	13	5	72
Portugal	64	2	82	12	161
Spain	391	9	291	56	747
Sweden	105	2	40	8	155
Switzerland	96	1	21	8	126
Turkey	65	11	42	11	129
United Kingdom	726	20	309	12	1 067
United States	8 518	43	1 432	524	10 517
Japan/RK	1 309	32	1 127	169	2 638
IEA Europe	4 699	97	2 151	321	7 268
US/Canada	9 089	49	1 535	576	11 249
Australia/NZ	338	9	78	77	501
Total IEA	15 435	188	4 891	1 143	21 656

ABBREVIATIONS

avg   average

bbls   barrels

Btu   British thermal unit

CAFE   Corporate Average Fuel Economy Standards (for US cars and light trucks)

CNG   compressed natural gas

LPG   liquid petroleum gas

CO_2 carbon dioxide

HOV   high-occupancy vehicle

km   kilometres

lbs   pounds

kg kilograms

kph   kilometres per hour

LDV   light-duty vehicle

MJ   megajoules

mph   miles per hour

NZ   New Zealand

PATP   "pay at the pump"

PAYD   "pay as you drive"

PKT   passenger-kilometres travelled

PMT   passenger-miles travelled

psi   pound per square inch

RK   Republic of Korea

ROW   rest of world

SUV   sport utility vehicle

TDM   travel demand management

UITP   International Union of Public Transport
         (http://www.uitp.com/home/index.cfm)

UK   United Kingdom

US   United States

VKT   vehicle-kilometres travelled

VMT   vehicle-miles travelled

REFERENCES

Barker William G. (1983), "Local experience", *Proceedings of the Conference on Energy Contingency Planning in Urban Areas*, Transportation Research Board Special Report 203.

Booz Allen Hamilton (2003), *ACT Transport Demand Elasticities Study*. Canberra Department of Urban Services, *www.actpla.act.gov.au/plandev/transport/ACTElasticityStudy_FinalReport.pdf*, April.

Bureau van Dijk (1992), *Évaluation de l'efficacité des mesures envisagées par les pouvoirs publics en cas de crise pétrolière*, Ministère des Affaires Économiques, Administration de l'Énergie, Belgium.

Cairns Sally, Carmen Hass-Klau and Phil Goodwin (1998), *Traffic Impact of Highway Capacity Reductions: Assessment of the Evidence*, Landor Publishing, London.

CEC (2003), California Energy Commission, *California State Fuel Efficient Tire Report, Volume II, Consultant Report*, *http://www.energy.ca.gov/reports/2003-01-31_600-03-001CRVOL2.PDF*.

Cambridge Systematics (1994), *The Effects of Land Use and Travel Demand Management Strategies on Commuting Behaviour*, Travel Model Improvement Program, US DOT.

Chatterjee Kiron and Glenn Lyons (2002), "Travel Behaviour of Car Users During the UK Fuel Emergency and Insights into Car Dependence", *Transport Lessons of the Fuel Tax Protests of 2000*, ed. by Glenn Lyons and Kiron Chaterjee, Ashgate: Aldershot.

Commission for Integrated Transport (2002a), *Achieving Best Value for Public Support of the Bus Industry, PART 1: Summary Report on the Modelling and Assessment of Seven Corridors*, *http://www.cfit.gov.uk/research/psbi/lek/chapter10/index.htm*.

Commission for Integrated Transport (2002b), *Fact sheet No.13: public subsidy for the bus industry*, *http://www.cfit.gov.uk/factsheets/13/*.

Dargay Joyce M. and Mark Hanly (1999), *Bus Fare Elasticities*, Report to the Department of the Environment, Transport and the Regions, ESRC Transport Studies Unit, UK.

De Jong Gerard, and Hugh Gunn (2001), "Recent evidence on car cost and time elasticities of travel demand in Europe", *Journal of Transport Economics and Policy*, 35 (2): 137-160.

Delucchi *et al.* (2000), *Electric and Gasoline Vehicle Lifecycle Cost and Energy-Use Model*, Institute of Transportation Studies (University of California, Davis) Paper UCD-ITS-RR-99-4.

DIW (1996), *The Efficiency of Measures to Reduce Petroleum Consumption in the Context of Supply Constraints*, German Institute for Economic Research, report commissioned by the German Federal Minister of the Economy.

ECMT/IEA (2005, in press), *Making Cars More Fuel Efficient: Technology and Policies for Real Improvements on the Road*, OECD/ECMT and IEA.

Eves David, James Quick, Paul Boulter and John Hickman (2002), The Effect of the Fuel Emergency on Sections of the English Motorway Network, in: *Transport Lessons of the Fuel Tax Protests of 2000*, ed. by Glenn Lyons and Kiron Chaterjee, Ashgate: Aldershot.

Ewing Reid and Robert Cervero (2002), "Travel and the Built Environment: A Synthesis", *Transportation Research Record*, #1780: 87-114.

Gillen David (1994), "Peak Pricing Strategies in Transportation, Utilities, and Telecommunications: Lessons for Road Pricing", *Curbing Gridlock*. Transportation Research Board special report 242: 115-151.

Gillespie T.D. (1992), *Fundamentals of Vehicle Dynamics*, Society of Automotive Engineers, Warrendale, Pennsylvania.

Giuliano Genevieve (1995), "*The Weakening Transportation-Land Use Connection*", *ACCESS*, Vol. 6, University of California Transportation Center, Spring, pp. 3-11. As cited by Litman, Todd at: *http://www.vtpi.org/ tdm/tdm15.htm*.

Goodwin Phil B. (1992), "A review of demand elasticities with special reference to short and long run effects of price changes", *Journal of Transport Economics and Policy*, 26 (2): 155-169.

Goodwin Phil, Joyce Dargay and Mark Hanly (2004), "Elasticities of road traffic and fuel consumption with respect to price and income", *Transport Reviews*, 24 (3): 275-292.

Graham, Daniel J. and Stephen Glaister (2002), "The demand for automobile fuel: A survey of elasticities", *Journal of Transport Economics and Policy*, 36 (1): 1-26.

Graham, Daniel J. and Stephen Glaister (2005, in press), "A review of road traffic demand elasticity estimates", *Transport Reviews*.

Hagler Bailly Inc. (1999), *Costs and Emissions Impacts of CMAQ Project Types*, prepared for US Environmental Protection Agency, Office of Policy.

Hartgen David T. and Alfred J. Neveu (1980), *The 1979 Energy Emergency: Who Conserved How Much?*, Preliminary Research Report 173, Research for Transportation Planning, New York State Department of Transportation.

Hensher David A. (1997), *Establishing a Fare Elasticity Regime for Urban Passenger Transport: Non-Concession Commuters*. Working Paper, ITS-WP-97-6, Institute of Transport Studies, University of Sydney, Sydney.

ICF Consulting (2003), *Greenhouse Gas Emissions Reductions from Current Transportation Programs*, prepared for EPA Office of Transportation and Air Quality (unpublished).

ICF Consulting and Imperial College Centre for Transport Studies (2003), *Cost-Benefit Analysis of Road Safety Improvements*, Final report for European Commission.

IEA (2001), *Saving Oil and Reducing CO_2 Emissions in Transport*, International Energy Agency, OECD/IEA, Paris.

IEA (2004), *Biofuels for Transport: An International Perspective*, International Energy Agency, OECD/IEA, Paris.

IEA (2004b), *World Energy Outlook 2004*, International Energy Agency, OECD/IEA, Paris.

IRTAD (2004), International Road Traffic and Accident Database, OECD, *http://www.bast.de/htdocs/fachthemen/irtad/*.

ITAC (2000), *Telework America 2000*, International Telework Association & Council, *http://www.telecommute.org/twa2000/research_results_key.shtml*.

Kain John F., Ross Gittell, Amrita Danier, Sanjay Daniel, Tsur Somerville and Liu Zhi (1992), *Increasing the Productivity of the Nation's Urban*

Transportation Infrastructure: Measures to Increase Transit Use and Car-pooling, US Department of Transportation, DOT-T-92-17.

Kuzmyak J. Richard (2001), *Cost-Effectiveness of Congestion Mitigation and Air Quality (CMAQ) Strategies*, Prepared for CMAQ Evaluation Committee, US Transportation Research Board.

LDA Consulting (2003), ESTC, and ICF Consulting; *TDM Strategy Assessments*; prepared for Southern California Association of Governments, Regional Transportation Demand Management Task Force.

Lee Martin E.H. (1983), *An International Review of Approaches to Demand Restraint in Transport Energy Contingencies*, Proceedings of the Conference on Energy Contingency Planning in Urban Areas, Transportation Research Board, Special Report 203.

Litman, Todd (2000), *Distance-based vehicle insurance: Feasibility, costs and benefits, comprehensive technical report*, Victoria Transport Policy Institute, *http://www.vtpi.org/dbvi_com.pdf.*

Litman Todd (2004), *Transit Price Elasticities and Cross-Elasticities*, Victoria Transport Policy Institute, Working Paper.

Luk James and Stephen Hepburn (1993), *New Review of Australian Travel Demand Elasticities.* Australian Road Research Board (Victoria), December 1993.

McDonald Noreen C. and Robert B. Noland (2001), "Simulated travel impacts of high-occupancy vehicle lane conversion alternatives", *Transportation Research Record*, 1765: 1-7.

Mendler C. (1993), "Equations for Estimating and Optimizing the Fuel Economy of Future Automobiles", SAE Technical Paper Series #932877, Society of Automotive Engineers, Warrendale, Pennsylvania.

Meyer, Michael D. (1999), "Demand management as an element of transportation policy: using carrots and sticks to influence travel behaviour", *Transportation Research A*, 33: 575-599.

MTC (1995), Metropolitan Transportation Commission *Impact of the Bay Area Commuter Check Program: Results of Employee Survey.* Oakland, California.

Nijkamp Peter and Gerard Pepping (1998), Meta-analysis for explaining the variance in public transit demand elasticities in Europe, *Journal of Transportation and Statistics*, 1 (1): 1-14.

Noland Robert B. and John W. Polak (2001), *Modelling and Assessment of HOV Lanes: A Review of Current Practice and Issues, Final Report*, submitted to the UK Dept. of Environment, Transport and the Regions.

Noland Robert B., John W. Polak and Gareth Arthur (2001), *An Assessment of Techniques for Modelling High-Occupancy Vehicle Lanes*, European Transport Conference.

Noland Robert B. and Lewison L. Lem (2002), "A Review of the Evidence for Induced Travel and Changes in Transportation and Environmental Policy in the United States and the United Kingdom", *Transportation Research D*, 7 (1) : 1-26.

Noland Robert B., John W. Polak, Michael G.H. Bell and Neil Thorpe (2003), "How Much Disruption to Activities Could Fuel Shortages Cause? The British Fuel Emergency of September 2000", *Transportation*, 30: 459-481.

ORNL (2003), *Transportation Energy Data Book, Edition 22*, Oak Ridge National Laboratory, *http://www-cta.ornl.gov/data/Download22.html*.

Pratt R.H. (1981), *Traveler Response to Transportation System Changes*, prepared for US Federal Highway Administration, DOT-FH-11-9579, July 1981.

Pratt Richard (1999), *Traveler Response to Transportation System Changes, Interim Handbook*. TCRP Web Document 12, DOT-FH-11-9579, National Academy of Science, *www4.nationalacademies.org/trb/crp.nsf/all+projects/tcrp+b-12*.

Pucher John (1997), Bicycling Boom in Germany: A Revival Engineered by Public Policy, *Transportation Quarterly*, 51 (4): 31-46.

Ross M. (1997), "Fuel Efficiency and the Physics of Automobiles", *Contemporary Physics*, 38: 381-394.

RSPA (1995), Research and Special Programs Administration, *TransitChek in the New York City and Philadelphia Areas*. Prepared for US Department of Transportation, Federal Transit Administration, October.

Shoup Donald C. (1997), "Evaluating the effects of cashing out employer-paid parking: eight case studies", *Transport Policy*, 4 (4): 201-216.

Southern California Rideshare (2003), *http://www.socalcommute.org*.

SCAG (1999) *State of the Commute Report*, South Coast Council of Governments, California, *http://www.scag.ca.gov/publications/pdf/SOC_1999.pdf.*

Thomas M. and M. Ross (1997), "Development of second-by-second fuel use and emissions models based on an early 1990s composite car", Society of Automotive Engineers Technical Paper Series #971010, Society of Automotive Engineers, Warrendale, Pennsylvania.

Thorpe Neil, Michael Bell, John Polak and Robert B. Noland (2002), "A Telephone Survey of Stated Travel Responses to Fuel Shortages", *Transport Lessons from the Fuel Tax Protests of 2000*, edited by Glenn Lyons and Kiron Chatterjee, Ashgate: Aldershot.

TRACE (1999), *The Elasticity Handbook: Elasticities for prototypical contexts (deliverable 5): costs of private road travel and their effects on demand, including short and longer term elasticities*, contract no. RO-97-SC.2035, Prepared for the European Commission Directorate-General for Transport.

Transit Cooperative Research Program (TCRP) (2003), *Land Use and Site Design, Traveller Response to Transportation System Change*s, Report 95: Chapter 15, Transportation Research Board, Washington, DC.

Transport for London (2003), *Congestion Charging: Six months on.*

UITP (1997), Urban Public Transit Statistics, Brussels.

UITP (2001), Millennium Cities Database for Sustainable Transport, Brussels.

Urban Transport Industry Commission (1994), *Inquiry Report No. 37, Volume 2: Appendices*, Australian Government Publishing Service, Melbourne.

US DOE (1994), *Energy, Emissions and Social Consequences of Telecommuting*, US Department of Energy, Office of Policy, Planning and Program Evaluation, DOE/PO-0026.

US EPA (1998), *Technical Methods for Analyzing Pricing Measures to Reduce Transportation Emissions*, Office of Policy, EPA 231-R-98-006.

VROM (2004), *Traffic Emissions Policy Document*, Netherlands Ministry of Housing, Spatial Planning and the Environment, *www.vrom.nl/international.*

WMCG (2002), *State of the Commute 2001: Survey Results from the Washington Metropolitan Region*, Washington Metropolitan Council of Governments, July.

The Online Bookshop

IEA PUBLICATIONS, 9, rue de la Fédération, 75739 PARIS Cedex 15
Printed in France by Corlet
(61 2005 22 1P1) ISBN 92-64-10-94-12 April 2005